Managing Psychological Factors in Information Systems Work:

An Orientation to Emotional Intelligence

Eugene Kaluzniacky
University of Winnipeg, Canada

 Information Science Publishing

Hershey • London • Melbourne • Singapore

Acquisition Editor:	Mehdi Khosrow-Pour
Senior Managing Editor:	Jan Travers
Managing Editor:	Amanda Appicello
Development Editor:	Michele Rossi
Copy Editor:	Maria Boyer
Typesetter:	Sara Reed
Cover Design:	Lisa Tosheff
Printed at:	Yurchak Printing Inc.

Published in the United States of America by
    Information Science Publishing (an imprint of Idea Group Inc.)
    701 E. Chocolate Avenue, Suite 200
    Hershey PA 17033
    Tel: 717-533-8845
    Fax: 717-533-8661
    E-mail: cust@idea-group.com
    Web site: http://www.idea-group.com

and in the United Kingdom by
    Information Science Publishing (an imprint of Idea Group Inc.)
    3 Henrietta Street
    Covent Garden
    London WC2E 8LU
    Tel: 44 20 7240 0856
    Fax: 44 20 7379 3313
    Web site: http://www.eurospan.co.uk

Library of Congress Cataloging-in-Publication Data

Kaluzniacky, Eugene.
  Managing psychological factors in information systems work : an orientation to emotional intelligence / Eugene Kaluzniacky.
    p. ; cm.
Includes bibliographical references and index.
  ISBN 1-59140-198-4 (cloth) -- ISBN 1-59140-290-5 (pbk.) -- ISBN 1-59140-199-2 (ebook)
  1. Emotional intelligence. 2. Computer programming--Psychological aspects.
  [DNLM: 1. Mental Health. 2. Occupational Health. 3. Emotions. 4. Information Services--organization & administration. 5. Personality. 6. Social Behavior. WA 495 K14m 2004] I. Title.
  BF576.K358 2004
  004'.01'9--dc22

                                   2003022610

British Cataloguing in Publication Data
A Cataloguing in Publication record for this book is available from the British Library.

All work contributed to this book is new, previously-unpublished material. The views expressed in this book are those of the authors, but not necessarily of the publisher.

To the memory of Maurice Senkiw

Managing Psychological Factors in Information Systems Work:

An Orientation to Emotional Intelligence

Table of Contents

Preface

With the expansion of the Internet and the resulting globalization of business activity, the scope of the influence of information technology (IT) has increased significantly. Many innovative business practices are being enabled by IT. The capacity for integration of information in numeric, text, voice, and video form will give rise to an even greater proliferation and impact of IT in the future.

Also, the information systems (IS) development profession has been maturing and IS has been recognized as a socio-technical endeavor for some time. For system developers, the need to communicate effectively with users and team members has been increasingly emphasized.

A recent (1993) survey of 192 human resources persons responsible for hiring new IS graduates in the Denver, Colorado, area revealed that, in addition to knowledge in applied computing and business, it was very important that a new IS hire be educated in: i) the ability to learn, ii) the ability to work in teams, iii) oral and written communication, iv) algebraic reasoning, and v) an orientation to health and wellness. In short, adaptability, communication, and stress management are seen as key skills for the IS professional. Yet, such skills are not developed through logic alone, but involve the "soft areas" of intuition, feelings, and senses.

Over 20 years ago, U.S. researchers Couger and Zawacki reported that, while IS professionals (systems analysts and programmers) had the lowest needs for social interaction on the job, they reported much higher "growth needs" than the other professionals surveyed. While, at the time, growth needs were largely understood as greater development of professional competencies, there now appears to be evidence that the IS development profession may be ready for a more wholistic approach to growth.

For example, a management scientist, in his book on IS management, has called for extending Maslow's hierarchy of needs beyond self-actualization to "self-donation" and has provided a concrete example of such a stage in the career of a systems analyst. An article in *Computerworld* has called for "emotional literacy among IS professionals" in the context of personality awareness. At a recent national convention of the Canadian Information Processing Society (CIPS), a keynote speaker proposed that "love" and not confrontation be the model for organizational communication, and received a standing ovation. A job advertisement for IT professionals within an insurance company in a prominent U.S. software center points out that this employer is interested in contributing to the employee's professional and personal life, and advises candidates to "listen to their inner voice." A prominent U.S. textbook author has referred, in the dedication of his text on IS for the Internetworked Enterprise. to "experiencing the Light within." In a recent Canadian survey on stress among IS professionals, the most frequently mentioned desired coping resource was "personal development seminars," closely followed by "conflict resolution seminars."

Also, the concept of "emotional intelligence" is being increasingly emphasized in management literature. It is being recognized that, while the traditional IQ (intelligence quotient) can help a person to *get* a job, it is the EQ (emotional quotient) that will allow the person to *keep* the job and to progress satisfactorily in his/her career.

Thus, the stage appears to be set for a preliminary attempt to address specific psychological factors as applied to the work of various IT professionals such as system and data analysts, programmers, project managers, help desk personnel, and also software engineers, telecommunications designers, and others. In this context, the term IT is considered to encompass a broader range of positions, whereas IS is more restricted to the activities of planning, analysis, design, development, and deployment of computerized business application systems.

The book is divided into two parts. In the first part, four psychological factors are considered: two personality type categories, cognitive style, and awareness of the inner self. Each area is introduced assuming no prior knowledge and then is related to situations from the work of information system developers. In the second part the focus is on *application* of the material presented in the first part to IT work.

Chapter I outlines in some detail the Myers-Briggs personality typing system. Prominent applications to IT that were made known in the form of published research studies and trade magazine articles are then highlighted. Since this system is well-structured and has been accepted in a significant number of managerial settings, most readers should be able to find the material both thorough and relevant. In this chapter questions for further research are also brought forth, and suggestions are offered for a person beginning his/her investigation of Myers-Briggs typing and IT work.

Chapter II follows up with a fairly comprehensive introduction to another personality typing system, the Enneagram. While Myers-Briggs focuses on *how* we function, this system offers insight as to *why* we do so and proposes an underlying emotion for each of its nine types. The Enneagram has also gained considerable acceptance in personal development circles and more recently in business and work. Each Enneagram type is outlined for a beginner's orientation, and then attempts are made to postulate strengths and shortcomings of each type in the course of IT work. Relationships between Myers-Briggs and the Enneagram are discussed, as is the complementarity of the two systems. As this is likely the first known attempt to link the Enneagram specifically to IT, references to existing research are replaced by considerable suggestions for future such endeavors.

Chapter III focuses on what may be considered a specific aspect of personality—cognitive style. The topic is introduced with the main distinction of sequential vs. intuitive (wholistic) approaches to perceiving. The reader is then introduced to a considerable number of significant research efforts in applying this factor to IT, since this area has indeed been researched more extensively. Cognitive styles are then expanded to include, specifically, creativity styles as well as learning styles, both very relevant to IS work. Again, research efforts are highlighted. Suggestions are also provided for the IT professional who is just becoming initiated with such perspectives.

Chapter IV is likely the most pioneering. It addresses the "deepest inner self" as a human component distinct from intellect, feelings, and body. Comprehensive, authoritative sources are used to present the material with a structured, analytic approach. In this fashion is also addressed the potential of human spirituality to empower IT workers with "psychological robustness." It is in this perspective that a structured, operational definition of "emotional intelligence" is developed, and the potential of such an intelligence in IS efforts is promoted considerably.

Part I concludes with Chapter IV. The intent for this part is, primarily, to provide a thorough introduction, particularly for a novice IS worker, to four prominent psychological factors. The reason for addressing these specific factors is largely the recognition that IS workers could benefit from greater awareness of themselves and others "as people," in order to develop "soft skills." An additional aim of Part I is to establish particular relevance of the addressed factors to IS work and to show how awareness of these factors can positively influence such work. Lastly, the chapters offer considerable material for consideration by MIS and interdisciplinary researchers.

Part II, consisting of Chapters V to VII, attempts to address application of the four major factors more specifically. Chapter V considers several specific areas of IT where the factors from Part I could make a difference, e.g., teamwork, end-user relationships, project management, interaction with organizational management, stress management, and human resource issues. This chapter, however, is an *initial* attempt to relate and to motivate. In time, after considerable feedback, a sequel book may be developed that would expand considerably on approaches of the chapter.

Chapter VI is written with a view towards developing an "emotionally intelligent IT organization." To this end, a growth stage model is presented, with five hypothesized stages as to how material from this book can be progressively and collectively adopted by an IT organization in the course of daily IT work. Chapter VII, which concludes the book, issues a "call to action" on the part of IS workers, their managers, higher executives, professional IT bodies, and academic researchers in facilitating the integration of psychological awareness into the IT skill set.

Thus, this book was written with two main objectives: i) to arouse awareness, among a broad spectrum of IT workers, of psychological factors that can contribute to their maturity and effectiveness at work; and ii) to catalyze specific, continuing efforts among both IT professionals and MIS academics, which would lead towards a deeper understanding and broader application of psychological issues in a variety of IT work situations. The book provides conceptual knowledge, reports on research findings, and presents both real and hypothesized anecdotes from the IT field. An attempt was also made to refer to both genders alternately when using pronouns; however, in most places, such pronouns should be easily interchangeable, reflecting no intended gender bias.

In this light, it is hoped that the book becomes more than "interesting reading" in one's spare time (if one has any). It is hoped that many IT professionals will consider this an introductory handbook to another dimension of their work, and will refer to it in an ongoing manner, learning from its insights, challenging its assumptions, and extending its boundaries. Likewise, it is hoped that MIS academics will be motivated by this book to conduct more research, so as to understand more profoundly the possibilities for increased effectiveness, efficiency, and fulfillment among IT professionals, resulting from increased psychological awareness.

However, communication is essential for such hopes to be realized. Questions, insights, and suggestions from all concerned must be publicized.

Ongoing development of emotional intelligence in IT must be a communal effort. To this end, a specific website related to this book is being planned. Communication provides motivation, and motivation begets creativity.

Just as medical doctors have realized that, in addition to scientific knowledge, bedside manners are essential to the profession, IT workers have also realized, at least in principle, that "soft skills" are increasingly required in the course of their work. However, specific applications have been sporadic and often incidental. May this modest effort motivate, both through what has been said and what has been omitted, awareness, discussion, and indeed definitive, coordinated and publicized action in expanding the boundaries of one of the most impactful professions of this decade.

Eugene Kaluzniacky
Winnipeg, Manitoba, Canada
November 2003

Acknowledgments

There are a number of individuals to whom I would like to express particular gratitude for assistance in the conceptualization, planning, and development of this book. I thank Vijay Kanabar, a former colleague, and David Erbach, my former department chair, for encouragement and motivation in exploring a new frontier. I also appreciate the assistance, encouragement, and hospitality of Cathal Brugha and Andrew Deegan of the Michael Smurfit Graduate School of Business in Dublin, Ireland, during my sabbatical. Three classes of MBA students at the school also provided valuable feedback on my preliminary presentations.

I thank Anita Chan for assistance in market planning and research, as well as Teresa Lesiuk, Janet Degelman and Theresa Jobateh for being valuable "sounding boards". I appreciate the clerical assistance of Celina Bibik and the excellent technical support from Fern Moran. As well, I would like to express heartfelt thanks to Chantal Antonie for her particular inspiration.

Lastly, I am indebted to the many IT professionals that I have met in the course of my teaching and research for their ideas, questions, and insight.

Part I:
Influential
Psychological Factors

When one accepts the vision that awareness of psychological dimensions within oneself and within one's work environment will indeed enhance the work of the IT professional, one must then wonder which psychological dimensions would be most worth investigating, at least initially.

This first and major part of the book introduces the reader to four psychological factors: two personality typing systems, cognitive styles, which then lead into specific areas of creativity and learning, and the transcendent reality of the "deepest inner self". Personality systems attempt to describe how (and perhaps to some degree why) we tend to behave and react in the course of our daily life. Cognitive considerations, more specifically, address how we think, create and learn. Clearly both these aspects of our psychological functioning can be hypothesized to play a role in the effectiveness of IT work. Furthermore, personality and cognition would likely be the first psychological factors that an IT "layperson" would want to consider for enhancing his/her work.

This part then goes a step further, in an attempt at generative creativity. It promotes the consideration of a more essential dimension of the human person, one considered in psychology that admits to a philosophic viewpoint. It shows a current trend in acknowledging and developing one's deepest core energy to which we can "connect" our thoughts and our actions. Then, a model of the "connected" human person is presented with a vision towards experiencing IT/IS work from a deeper and more empowered inner awareness.

The aim was to choose psychological factors that would provide relevance, complementarity and balance. Each psychological factor is discussed with enough thoroughness to initiate those marginally aware. Thus, each chapter can be considered as introductory text material to a particular psychological dimension as well as an effort to initiate a vision of the applicability of the dimension to IT/IS work. It may indeed be that readers with different personalities may derive most benefit from different chapters. Even then, resulting discussions could become fruitful and stimulating for those willing to "experiment" and grow.

Chapter I

The Myers-Briggs Personality Types

OVERVIEW

Just over seven years ago, I was on a flight home from the United States. Seated next to me was an IS professional coming to give training sessions on project management software that had recently been purchased by our local hydro-electric utility. Her training included an undergraduate degree in Business (IS major) and a Master's degree in Management Science. When I asked her if this educational background was adequate for her current job, she replied, almost immediately, that one area that was never covered in her studies was that she "would have to work with such different personalities." At that point, she had no idea of my occupation, specific interest in IT personalities, or the fact that three days later I would giving a workshop on personality types in IS work to a convention of IS professionals.

Personality can be defined as "a complex set of relatively stable behavioral and emotional characteristics" of a person (Hohmann, 1997). It refers, essentially, to how a person functions in life. Most of us, even without any training in this area, will recognize that the world consists of people of different types. We notice that people of different types will often react differently to the same situation. But when considering personality awareness as a desirable "soft skill" for the IT profession, many of us may stop to wonder. Yes, we may say, there are different types of airline pilots, athletes, and molecular biologists. But, is this an issue that is closely connected to work? It depends on the factors of which the work consists. Most would agree that personality relates to communication, learning style, and to what one finds stressful. Yet ability to learn, to communicate, to reason, and to maintain one's health are key competencies for which recruiters of new IS graduates are looking. Thus, it is postulated that the IT area, and system development (IS) in particular, can benefit significantly from awareness of particular characteristics of different personality types.

The Myers-Briggs Personality Type approach to classifying personalities has been widely accepted and applied in a diversity of fields such as social work, counseling, career planning, and management. It assesses four different dimensions of a person:

1. *Introversion/Extraversion:* relates to *how a person is oriented,* where he/she focuses more easily; within oneself or on other people and the surrounding environment. This dimension is coded I or E respectively.
2. *Intuition/Sensing:* relates to two different *ways of perceiving,* of taking in information. An intuitive person focuses on new possibilities, hidden meanings, and perceived patterns. A sensing person focuses on the real, tangible, and factual aspects. Thus a sensing person can be described as being more practical, whereas an intuitive is more imaginary. This dimension is coded N for Intuitive and S for Sensing.
3. *Thinking/Feeling:* relates to *how a person comes to conclusions,* how a person normally prefers to make judgments. A thinking person employs logical analysis, using objective and impersonal criteria to make decisions. A feeling person, on the other hand, uses person-centered values and motives to make decisions. This dimension is coded T for Thinking and F for Feeling.
4. *Judging/Perceiving:* relates to two essential attitudes of dealing with one's environment. A judging person prefers to make judgments, or come to conclusions about what one encounters in one's outer environment. A perceiving person prefers to notice one's outer environment, while not coming to conclusions or judgments about it. This dimension is coded J for Judging and P for Perceiving.

Thus, we see that the Myers-Briggs personality classification system identifies personality according to four dimensions. Since there are two possibilities for each dimension, there are 16 different Myers-Briggs personality types. Elaborating on the four dimensions, Extraversion/Introversion refers to where a person gets most psychological energy. Sensing/Intuition points out to what a person pays most attention. Thinking/Feeling shows how a person prefers to make decisions, and Judging/Perceiving relates a preferred attitude to life.

An extravert generally prefers to draw energy from the outside: from interaction with other people, activities, or things. An introvert prefers to draw energy from ideas, emotions, or impressions within.

A sensing person takes in information through five senses and notices more of what is actual, tangible, factual. An intuitive person takes in information mostly through a "sixth sense," and thus focuses on hidden meanings and new possibilities.

A thinking person prefers to decide in a logical, objective way, whereas a feeling person prefers to decide in a personal, valued-oriented way.

Finally, a judging person prefers to live a planned, organized, structured life, and enjoys results from a task more than the process that leads to the results. A perceiving person prefers to live an open-ended, spontaneous life and enjoys the processes of work tasks more than results.

It is important to appreciate that what we are assessing on each dimension is a person's *preference*. While we all can function to some degree on both sides of each dimension, it will be easier and more natural to function on one side, and we will thus prefer to function on that side more often. To illustrate, a thinking type certainly has feelings and may have a fairly rich emotional life. However, this type tends to come to conclusions in a logical, objective way. A feeling type may have significant capacity for abstract logic, but prefers to decide on the basis of personal values and feelings.

One may ask, at this point, why are these particular four dimensions involved in classifying personalities? The first three dimensions were identified by renowned psychiatrist Carl Gustav Jung; the last dimension was introduced at a later time. Over nearly a century, it has been observed that indeed the four dimensions of energizing, attending, deciding, and living attitude account significantly for the differences in the human behavioral characteristics that we call "personality."

If we first consider the fourth dimension, living attitude, a person with a judging preference will be more often involved in deciding (the third dimension) than attending (the second dimension). Conversely, a perceiving person will prefer to pay attention (the second dimension) rather than to draw immediate conclusions (the third dimension). Thus, starting at the fourth dimension of personality leads us to consider the third and second dimensions. Furthermore, in addition to having a judging or perceiving living attitude, personalities differ on the dimension of deriving energy, i.e., extraversion/introversion (the first dimension).

It is important, at this point, to admit and to emphasize that people do differ in more ways than the above four dimensions. Ultimately, each individual is unique and unrepeatable. However, the Myers-Briggs system classifies personalities into 16 types that account for differences in people on some key personality characteristics. Moreover, this system has been found to be comprehensive enough so as to provide useful insights, through comparison and contrast, for the personal and professional lives of a large number of people throughout the world. Thus, it is not unreasonable to believe that an orientation in the Myers-Briggs personality classification system can indeed make a valuable contribution to the IT profession.

HISTORICAL BACKGROUND

The Myers-Briggs system was largely developed in the first part of the twentieth century in the United States by a mother-daughter team. In 1917, a largely self-educated woman, Katherine C. Briggs, began studying differences in people through reading biographies. A while later, she saw that many of her observations were corroborated by Carl G. Jung in his book, *Psychological Types* (1961).

Thus, between 1923 and 1941, Katherine Briggs and her daughter Isabel Briggs Myers studied Jungian theory in more depth and observed personalities in terms of the theory. In 1941, a decision was made to create a personality questionnaire, an indicator, so that Jungian theory could be applied, particularly in the war effort. In 1942-44, early forms of the Myers-Briggs Type Indicator (MBTI) were developed. The indicator questionnaire was based on theory and observation. Briggs and Myers had added the Judging/Perceiving dimension to Jung's Extraversion/Introversion, Sensing/Intuition, and Thinking/Feeling.

It is noteworthy to point out that the questionnaire which attempts to determine one's Myers-Briggs type is called a "personality indicator" and not a "personality test," since one cannot pass or fail the questionnaire and there is no minimally desirable score.

After the Second World War, more extensive research was undertaken on MBTI data. Specific samples, such as those of medical and nursing students, were analyzed. In 1956, Educational Testing Services (ETS) became the publisher of the MBTI as a research instrument. In 1956-62, new, refined forms of the Indicator were published, and in 1962-69, isolated researchers and clinicians became acquainted with the Indicator and found it useful. In 1971 a Typology Laboratory was created at the University of Florida, and in 1975 Consulting Psychologists Press became the publisher of the MBTI, which then became widely available to the psychological community for the first time. In 1979, the Association for Psychological Type (APT) was formed. It is now the "overseer" of the MBTI and conducts training workshops at introductory and advanced levels in order to develop qualified MBTI type counselors.

Isabel Briggs Myers died in May 1980. Her celebrated book, *Gifts Differing* (1995), explains the foundations of the Myers-Briggs personality types, named after the daughter and mother who classified them. For the past 20 years, millions of people have taken the Myers-Briggs Type Indicator in a variety of countries. The Indicator is normally taken through a person who is qualified to administer and interpret it.

In the 1970s, another classic book appeared on the Myers-Briggs personalities, *Please Understand Me* (1978) by David Keirsey and Marilyn Bates. This book contains a shortened adaptation of the Myers-Briggs Indicator called the "Keirsey Sorter," which can give the reader a reasonable idea of how he might come out on the official MBTI. In addition, Keirsey collapses the 16 types into four *temperaments*: NT (Rational), NF (Idealist), SJ (Guardian), and SP (Artisan).

Thus, each MBTI type belongs to exactly one Keirsey temperament. Each temperament is thus matched with four Myers-Briggs types, and allows applications on a more aggregated basis. The Keirsey Sorter, in addition to being in the book mentioned above, is also available on the World Wide Web, free of charge.

MBTI AT WORK

There is considerable documentation outlining general work characteristics of each Myers-Briggs dimension. Many of these modes of operation can have significant relevance to system development work. Some highlights from this area are now presented.

Extraverts like variety and action, and are often impatient with long, slow jobs. They are more interested in how other people do a job similar to them. They often act quickly, sometimes without much thought, and communicate freely. They like to develop ideas with others.

Introverts, of course, like quiet concentration. They tend not to mind working on one project for a long time, without interruption, and like to think before they act. They may have problems communicating and develop ideas alone.

Sensing individuals dislike new problems unless there are standard ways to solve them. They are not likely to initiate or promote change unless absolutely necessary. They enjoy using skills they have already learned more than learning new skills. Introverted sensors are especially patient with routine details (particularly ISJ). They also tend to be good at precise work and are very practical.

Intuitives, on the other hand, dislike doing the same thing repeatedly (especially NPs). They enjoy learning new skills. They tend to work in short bursts of energy, powered by enthusiasm with slack periods in between (especially ENs). They tend to be impatient with fine details, preferring concepts to details, and frequently make error of fact. Intuitives, also, are more prone to inspiration than sensors.

Thinking types like analysis and putting things into logical order. They are more likely to get along without interpersonal harmony, and respond more easily to people's thoughts than feelings. Thus, they tend to decide impersonally, often not weighing various effects of their decisions on people's feelings. They (especially ITs) do not show emotion readily and are often uncomfortable dealing with feelings. Thinkers tend to be firm-minded, as opposed to sympathetic, and need to be treated fairly. They choose truthfulness over tactfulness.

Feeling types are more aware of their own feelings and those of others. They like harmony in the workplace. When making decisions, they tend to be influenced by their personal likes and dislikes, as well as those of others, and dislike telling people unpleasant things. They tend to be more sympathetic than firm and may need praise in addition to fairness. Feelers fit quite comfortably in the position of team builders.

Judging (structured) types work best when they can plan their work and follow the plan. Thus, they may dislike interrupting the current project for a more urgent one (especially SJs). They may not notice new things that have arisen and need attention because this would disrupt their schedule. They like to get things settled and finished, and may decide too quickly (especially EJs). Judgers generally focus on the result of the work much more than on the work process itself. They would thus feel much more upset if a project was cancelled or had failed (especially FJs).

Perceiving (open-minded) types adapt much more readily to abrupt changes in schedule. In fact, schedules often "cramp their style" and may drain their energy. They tolerate interruptions quite well and even prefer spontaneity. Since they perceive people, situations, or things without drawing conclusions, they may have trouble making decisions because they want to remain open to new information. They may start too many projects without finishing them (especially ENPs), and may postpone unpleasant jobs.

MBTI IN IT

The MBTI had made its way into IS over 15 years ago. In "The DP Psyche" (*Datamation,* 1998), Michael Lyons reports on an international survey of personalities of more than 1,000 professionals employed by over 100 different companies. About one-third of those surveyed were employed as programmers or analysts, and about 20% were in IS management. Table 1 shows the breakdown along the four MBTI dimensions. We notice twice as

Table 1: Breakdown by Personality Dimension of 1,229 System Development Professionals (Lyons' Datamation Study)

Introverts:	67%	Intuitives:	54%
Extraverts:	33%	Sensors:	46%
Thinkers:	81%	Judgers:	81%
Feelers:	19%	Perceivers:	19%

Table 2: Most and Least Frequent Types in IS (Lyons' Datamation Study)

Most Frequent:	ISTJ	22.6%
	INTJ	15.5%
	INTP	12.1%
	ESTJ	9.3%
	ENTJ	8.4%
Least Frequent	ISFP	1.5%
	ESFP	1.5%
	ESFJ	1.0%

many introverts as extraverts, slightly more intuitives than sensing people, a very high percentage (80-90%) of thinking types, and a two-to-one ratio of judging to perceiving types.

The article sheds significant, albeit introductory, light on the distribution of MBTI characteristics in a large, dispersed sample of IS workers. A distribution of actual types is also presented.

Table 2 shows the percentage of those surveyed in the four most frequent and three least frequent types. The results shown lead one to ask two important questions: Does the observed distribution of IS types correlate with Myers-Briggs type theory? Does this distribution vary with other samples?

We now examine the main characteristics of each of the four dimensions as they relate to IS work.

Extraversion/Introversion

While there clearly is room for both extraverts and introverts in systems development, it is not surprising to see a 2:1 ratio of introverts:extraverts. Tasks such as detailed data modeling, coding, quality assurance testing, and network design can lend themselves quite well to preferred introversion. However, extraverts can feel especially at home in requirements determination, Joint Application Development, presentation to users/senior management, user training, and help desk activities, for example.

Sensing/Intuition

The clear distinction here is "reality thinking" vs. "possibility thinking." A considerable amount of system development activity definitely fits with and appeals to the practical, details-and-facts-oriented sensing mentality. Much of actual technology is practical; activities such as system installation, detailed telecommunication design, physical data modeling, as well as programming, testing, activity scheduling, and detailed documentation would appeal to and energize the sensing person.

However, there are certainly more conceptual aspects to system development, some more structured and others more open-ended. Activities such as system planning, high-level business and data modeling, object modeling, and political "positioning" would be much more in the realm of intuitive types. Since there is considerable opportunity for both sensors and intuitives to find IT work appealing, in Lyons' survey we find the most balance along sensing/intuitive lines (46% S to 54% N).

Thinking/Feeling

We recall that while both types certainly think and feel, the thinking types prefer to decide with logical analysis while feelers tend to base decisions on personal values and feelings. Considerable IS development activity, no doubt, involves the thinking function, whether it be practical thinking (as in telecommunication design or testing) or conceptual thinking (object modeling, system planning). Often, the thinking must be structured and yield specific deliverables that can execute on specific machines. But, how can feeling types find a home in IS work? Since they place considerable focus on harmony, feelers can be particularly sought after as group/team leaders, high-level business modelers, or analysts, where considerable effective interaction with non-IT staff is essential. Feelers may become prominent IS "politicians" who can forge effective relationships with others in organizations. They can also contribute innovatively and effectively in development of training materials and in the training process itself. As systems move towards integration of a variety of communication modes through multimedia and Internet access, the contribution of artistically minded feelers will be increasingly desirable. It is worth pointing out that an "F" is a person who *prefers* to decide from personal values, but he/she may be more capable or less so in exercising the logical thinking function when it is called for. Most Fs in IT, however, would need to possess a well-developed capacity for thinking. Lyons' study showed an overwhelming proportion (80.9%) of thinking types. Later, we will see evidence that such significant dominance may slowly be changing.

Judging/Perceiving

This dimension relates to the need for order, structure, and closure in one's life and work. Computing itself is largely structured with emphasis on precision. Thus procedural language programming, for example, would be ideal for a Judging orientation as would be detailed telecommunication design. Yet, there certainly are activities in the development and mainte-nance of systems where too much structure and predictability would not be desirable. System planning and brainstorming, for example, thrive on flexibility and spontaneity. Business and data modeling for a new system also mandate adaptability and flexibility. Maintenance and help desk work is often unpredictable and varied. Lyons' study shows an almost 2:1 ratio of Judging to Perceiving types. This reflects a large reliance on structure, but admits open-endedness as a preference in one-third of the sample.

We thus see that each part of each Myers-Briggs dimension has a role to play in system development work. In this field, it is beneficial if the individual is flexible and has developed considerable strength on the sides that are not his/her preference. Also, it is very desirable to have a variety of personalities in various facets of system development.

Type Occurrences in IS Work

As we have seen in Table 1, Michael Lyons has presented statistics on frequencies relating to each MBTI dimension. Yet, we need to ask ourselves how representative are these numbers, and how do distributions for IS workers compare to those for the general population?

Table 3 shows data by personality dimension from several sources. The first line (Society) shows the distribution for each dimension as observed in North American society. The second line (Datamation) refers to the figures in Lyons' study. The remaining lines provide figures from several studies of my own. The third line (Cont Ed.) provides statistics from a survey of 28 IS professionals in an Information Resources Management Continuing Education course in the early '90s. The fourth and fifth lines (Org. 1 and Org. 2) pertain to studies of IS professionals in two Winnipeg organizations. Org. 1 was a health administration body and Org. 2 an investment firm. The number of subjects in each study was approximately 30. The last line (Project) refers to approximately 60 students in a final-year undergraduate system development project course. We now note further insight for each personality dimension.

Table 3: Breakdown by Personality Dimension from Several Sources

	Introvert/ Extravert	Sensing/ Intuitive	Thinking/ Feeling	Judging/ Perceiving
Society	25-75%	75-25%	50-50%	55-45%
Datamation	67-33	46-54	81-19	66-34
Cont. Ed.	66-34	47-53	73-27	73-27
Org. 1	60-40	67-33	80-20	73-27
Org. 2	73-27	42-58	66-34	55-45
Project (UofW)	68-32	73-27	86-14	80-20

Extravert/Introvert

While in the general society 75% of the people are extraverted, it is not surprising to see significantly more introverts in the IS profession. What is quite surprising, however, is that in four of the five cases, the ratio of introverts to extraverts is very nearly 2:1. This comes from entirely different samples in different locations and with different sample sizes. Although such results are, at best, exploratory, they seem to point to a possible homogeneity across the profession, with one extravert for every two introverts. Thus, there seems to be a consistent and sizeable presence of extraverts in IS, although they are in the minority.

Sensing/Intuitive

In general society, 75% of the people prefer sensing (practicality) to intuition. Lyons reported 54% of intuitives and my Continuing Education sample showed 53%, again a remarkable similarity. The second organization that I studied showed a similar 58% intuitives. The first organization, however, had a markedly lower percentage of intuitives.

In Org. 1, systems analysts were recruited almost exclusively from programmer ranks, whereas in Org. 2, some of the systems analysts had come from user departments. This quite possibly relates to a higher proportion of intuitives in the second organization. In the university Project course, the sensing/intuitive break mirrors the general population. However, as information systems become more influential, and their scope and content broaden, we can certainly see major opportunities for the intuitives, who are "possibility thinkers."

Thinking/Feeling

Although the two preferences are split evenly in general society, it is definitely not surprising to see more thinkers in IS. In most cases, the data show 70-80% thinkers. We must recall that a thinker *prefers* to come to conclusions with rational, logical analysis, while a feeler is more prone to involve personal values and a need for harmony. Thus, this does not imply that feeling types are not *capable* of logical reasoning; for them, however, this is not the ultimate.

Among my results, Org. 2, where some of the analysts have come from user ranks, shows more feeling types than Org. 1, where the analysts had been largely promoted from programmer ranks. Org. 1 had 80% thinkers, parallel to the Datamation study with 81%. In the university Project course, 86% of the students were thinkers, since a rigorous complement of programming courses (in Pascal, COBOL, and C) was required to enter into the Project.

Judging/Perceiving

In general North American society, there is a slight bias of 55-45% towards judging types. Lyons' sample showed 66% judging types, but in the second organization, the breakdown resembled that of the population. The Continuing Education sample, as that of the first organization, showed even more judging types, each 73% to 27%, while the Project course had an overwhelming 80% judgers. Traditionally, when IS was equated with programming, the field would be much more attractive to judging types who thrived on structure and closure. However, the strength of the perceiving individual is flexibility and a tendency not to decide too soon. Such traits are clearly desirable in the IS profession of the future.

In all, the breakdown clearly points to introversion, thinking, and judging as most frequent traits of IS personnel. This certainly can be expected. However, three of the five samples showed a slightly higher percentage of intuitives than sensing types. At first thought, one would expect more sensing types in IS, since much of the work is factual and could be considered "tangible." With the increasing roles of intuitives in the systems of the future, the breakdown deserves further investigation.

Distributions of Individual Types

We have just examined various samples of data on IT professionals, dimension by dimension. Now, we look at actual distribution among the 16 Myers-Briggs types.

Firstly, we see the concept of a Type Table, as shown in Table 4. This is simply an ordered collection of squares, one for each type. Such a table for MBTI, as defined by the Association for Psychological Type, has each type always in the same position. For example, ISTJ is always in the upper left and ESTJ in the lower left. Thus, this table can be divided into four *quadrants*: IS—upper left, IN—upper right, ES—lower left, EN—lower right. Table 5 also contains descriptions of each type. This figure first appeared in *The American Programmer* in 1990. Table 6 now shows a type table with three separate personality distributions. In each square, the top figure refers to Lyons' study that was referenced earlier, the middle figure is from a study of 656 Australian IS employees as reported by Thomsett (1990), and the bottom figure relates to my sample of 66 Project course students at the University of Winnipeg. We now examine significant features in this figure.

Table 4: Basic Type Table

ISTJ	ISFJ	INFJ	INTJ
ISTP	ISFP	INFP	INTP
ESTP	ESFP	ENFP	ENTP
ESTJ	ESFJ	ENFJ	ENTJ

Table 5: Type Table (according to American Programmer, 1990)

ISTJ Serious, quiet, earn success by concentration and thoroughness. Practical, orderly, matter of fact, logical, realistic, and dependable. Take responsibility	**ISFJ** Quiet, friendly, responsible and conscientious. Work devotedly to meet their obligations. Thorough, painstaking, accurate. Loyal considerate.	**INFJ** Succeed by perseverance, originality, and desire to do whatever is needed, wanted. Quietly forceful; concerned for others. Respected for their firm principles.	**INTJ** Usually have original minds and great drive for their own ideas and purposes. Skeptical, critical, independent, determined, often stubborn.
ISTP Cool onlookers – quiet, reserved, and analytical. Usually interested in impersonal principles, how and why mechanical things work. Flashes of original humor.	**ISFP** Retiring, quietly friendly, sensitive, kind, modest about their abilities. Shun disagreements. Often relaxed about getting things done.	**INFP** Care about learning, ideas, language, and independent projects of their own. Tend to undertake too much, then somehow get it done. Friendly but often too absorbed.	**INTP** Quiet, reserved, impersonal. Enjoy theoretical or scientific subjects. Usually interested mainly in ideas, little liking for parties or small talk. Sharply defined interests.
ESTP Matter-of-fact, do not worry or hurry, enjoy whatever comes along. May be a bit blunt or insensitive. Best with real things that can be taken apart or put together.	**ESFP** Outgoing, easygoing, accepting, friendly; make things more fun for others by their enjoyment. Like sports, making things. Find remembering facts easier than mastering theories.	**ENFP** Warmly enthusiastic, high-spirited, ingenious, imaginative. Able to do almost anything that interests them. Quick with a solution and to help with a problem.	**ENTP** Quick ingenious, good at many things. May argue either side of a question for fun. Resourceful in solving challenging problems but may neglect routine assignments.

Table 5: Type Table (according to American Programmer, 1990), Cont.

ESTJ Practical, realistic, matter-of-fact, with a natural head for business or mechanics. Not interested in subjects they see no use for. Like to organize and run activities.	**ESFJ** Warm-hearted, talkative, popular, conscientious, born co-operators. Need harmony. Work best with encouragement. Little interest in abstract thinking or technical subjects.	**ENFJ** Responsive and responsible. Generally feel real concern for what others think or want. Sociable, popular. Sensitive to praise and criticism.	**ENTJ** Hearty, frank, decisive leaders. Usually good in anything that requires reasoning and intelligent talk. May sometimes be more positive than their experience in an area warrants.

The *most frequently occurring type in all these samples is ISTJ* (introverted, practical, thinking, and structured). The type is described further in Table 5. Such a type is particularly suited to traditional, procedural programming. Both the Australian and Winnipeg samples show ESTJ in second place as to frequency and INTJ in third place. In fact, these top three types account for 70.1% and 70.6% of the sample in the two samples, respectively. This, again, is quite remarkable considering the differences in country, age, and sample size. The Lyons study shows a more dispersed sample. While ISTJ also comes first, INTJ comes second, followed by INTP and ESTJ. Here, the top three types account for 50.6% of the sample. In the Australian, Winnipeg, and Lyons samples, 77.8%, 78.1%, and 73.9% of the sample is accounted for by the five most frequently occurring types.

Even though three samples cannot claim to represent the population of IT professionals, we do note meaningful insights. Firstly, there appears to be an influence of type on the choice of an IS career—not all 16 types are represented equally among IS workers. Secondly, types whose characteristics match the kinds of tasks required of IS workers are represented more frequently. This may lead us to develop some trust, amid curiosity, in the possibility of MBTI indeed contributing something of value to the IS profession.

Table 6: Type Table from Three Sources (Lyons, 656 Australian Employees, 66 U of W Project Students)

Studies	ISTJ	ISFJ	INFJ	INTJ
Lyons	22.6%	3.9%	2.7%	15.5%
Australia	38%	5.2%	2.9%	6.5%
U of W	39%	4.5%	-	12.1%
	ISTP	ISFP	INFP	INTP
Lyons	5.2%	1.5%	3.4%	12.1%
Australia	0.6%	0.6%	0.5%	0.6%
U of W	6%	-	4.5%	1.5%
	ESTP	ESFP	ENFP	ENTP
Lyons	2.1%	1.5%	3.4%	5.6%
Australia	0.6%	1.5%	1.5%	1.7%
U of W	1.5%	-	1.5%	4.5%
	ESTJ	ESFJ	ENFJ	ENTJ
Lyons	9.3%	1.0%	2.4%	8.4%
Australia	25%	4.9%	3.8%	6%
U of W	18%	3%	-	3%

ORDER OF PREFERENCE

In order to see a clearer relationship between Myers-Briggs types and systems development activities, we examine some relationships among the second (S,N) and third (T,F) MBTI preferences. These preferences are referred to as *functions*. For each type, one of these four functions has the role of *dominant* function. It "takes the lead." The dominant function is used in the outside world by extraverts and in the inside world by introverts. The other function, that is, the other of the middle two letters for a given type, is called the *auxiliary* function; it supports the dominant function. Moreover, the dominant and auxiliary functions are used in opposite worlds. If the dominant function is extraverted, the auxiliary one will be introverted, and vice versa. The third or *tertiary* function is the opposite of the auxiliary. and the fourth or *inferior* function is the opposite of the dominant. Of the four functions (four possible middle letters), the dominant is the most preferred and the inferior is least preferred. Also, the letters for the dominant and secondary function appear (in the second and third positions) among the four letters of a particular type, the letters for the tertiary and inferior function will not appear in the type code. We also need to keep in mind that there are two possible perceiving functions: sensing and intuition, and two possible judging functions, thinking and feeling.

To shed a bit more light, in ISTJ we know that the possibilities for dominant and secondary functions will be S and T. For ENFP, N and F will be possibilities for dominant and secondary functions, since they are the middle two letters for this type. But how can we pinpoint exactly which of the two possibilities is actually the dominant function? The following guidelines help us.

For both extraverts and introverts, a P as a fourth letter means that the extraverted function is a *perceiving* one. We know that possible perceiving functions are sensing and intuition, and possible judging functions and thinking and feeling. We also know that the dominant function is extraverted by extraverts and introverted by introverts.

I shall now use three examples to illustrate how, given a particular MBTI type, the *order of preference* among the four functions can be derived.

ISTP: Here, the last letter being P, we know that the extraverted function is the perceiving one, in this case, sensing. However, since this is an *intro*verted type, the *extra*verted function will not be dominant but auxiliary. Thus, sensing is the auxiliary function here. Also, the dominant and auxiliary functions are to be chosen from among the {S,T} pair. Since it has already been determined that S is auxiliary, T is therefore the dominant function of the ISTP. The tertiary function is the opposite of the auxiliary, in this case, N, and the inferior function is the opposite of the dominant, in this case, F. Thus, we now have an *order of preferences* for this type, that is, an ordered listing of the four functions (the order in which each "kicks-in"): thinking, sensing, intuition, feeling.

INTJ: If the last letter is a J, this means that the extraverted function is the *judging* one. Here the judging function is T. If T is extraverted (shown to others), then the other of the middle two letters, N, is introverted. Since the first letter is I, the dominant function is introverted, thus the dominant function must be N, and the auxiliary function T. As a corollary, the tertiary function is F (opposite of auxiliary) and the inferior function is S (opposite of the dominant N). Note, as indicated previously, that the tertiary and in-ferior function code letters do *not* appear in the type code (F and S do not appear in INTJ). The order of preference for this type is therefore intuition, thinking, feeling, sensing.

ESFP: The last letter being P indicates that the extraverted function is the perceiving one, in this case S. Since this is an *extraverted* type, the dominant function is the extraverted function, thus the dominant function is S, making F the auxiliary function. As a result, T is tertiary and N is inferior. Here, the order of preference is sensing, feeling, thinking, intuition.

Thus, *each of the 16 types has an order of preference*, that is, a list of the four functions in a specific order. This order is different for every type.

Table 7 shows the order of preference to the 16 types. Such insight can indeed be valuable for system development since *different types will "drill through" a specific problem in different functional order*. Thus, they will

Table 7: Order of Preference for Each Type (from Hirsh and Kummerow)

ISTJ	**ISFJ**	**INFJ**	**INTJ**
1. Sensing (I)	1. Sensing (I)	1. Intuition (I)	1. Intuition (I)
2. Thinking (E)	2. Feeling (E)	2. Feeling (E)	2. Thinking (E)
3. Feeling (E or I)	3. Thinking (E or I)	3. Thinking (E or I)	3. Feeling (E or I)
4. Intuition (E)	4. Intuition (E)	4. Sensing (E)	4. Sensing (E)
ISTP	**ISFP**	**INFP**	**INTP**
1. Thinking (I)	1. Feeling (I)	1. Feeling (I)	1. Thinking (I)
2. Sensing (E)	2. Sensing (E)	2. Intuition (E)	2. Intuition (E)
3. Intuition (E or I)	3. Intuition (E or I)	3. Sensing (E or I)	3. Sensing (E or I)
4. Feeling (E)	4. Thinking (E)	4. Thinking (E)	4. Feeling (E)
ESTP	**ESFP**	**ENFP**	**ENTP**
1. Sensing (E)	1. Sensing (E)	1. Intuition (E)	1. Intuition (E)
2. Thinking (I)	2. Feeling (I)	2. Feeling (I)	2. Thinking (I)
3. Feeling (E or I)	3. Thinking (E or I)	3. Thinking (E or I)	3. Feeling (E or I)
4. Intuition (I)	4. Intuition (I)	4. Sensing (I)	4. Sensing (I)
ESTJ	**ESFJ**	**ENFJ**	**ENTJ**
1. Thinking (E)	1. Feeling (E)	1. Feeling (E)	1. Thinking (E)
2. Sensing (I)	2. Sensing (I)	2. Intuition (I)	2. Intuition (I)
3. Intuition (E or I)	3. Intuition (E or I)	3. Sensing (E or I)	3. Sensing (E or I)
4. Feeling (I)	4. Thinking (I)	4. Thinking (I)	4. Feeling (I)

ask different questions at different times in their deliberations. For example, we can assume two analysts, an ISTJ and an ENFP, are working on requirements determination for a new sales order system. Since the letters in these two types are the opposite of each other on every dimension, their order of preference, as seen in Table 7, will be exactly opposite.

The ISTJ will likely notice facts and details first: how sales order data are captured now, what data items are captured, where and how the data are stored, and what output is produced from the stored data. He will thus likely look at the practical, factual details of how the new system will operate, i.e., he will tend to focus on details of input, processing, output, and connectivity. He will focus on the facts (a sensing function) and then structure the facts in his mind (a thinking function). Since many ISTJs do not have well-developed feelings and intuition, only "basic" feelings may be felt at this stage, such as a sense of urgency and perhaps doubt if the deadlines can be met. The ISTJ would not likely use his inferior function, intuition, to any significant degree. The ISTJ's sense of system requirements would tend towards the practical and detailed, with possibly early questions about practicality and feasibility of some of the options under consideration.

The ENFP analyst (who may have risen to her current analyst position from the user side) will likely engage intuition first. She will search for possibilities of, for example, data entry at the customer site and connectivity through wireless technology. She may use a combination of intuition and feelings to perceive that this system is important for the marketing manager in order for her to receive a promotion six months hence. She may then intuit that the marketing manager may be pressing for options that, while looking impressive, may not justify the cost versus the benefit. She will be guided by the intuition-feelings combination to compliment the marketing manager genuinely on her vision and drive to improve the standards of marketing practice. Simultaneously, she may express genuinely, but convincingly, that options of online, on-site comparisons with other industry segments country-wide will be better off as enhancements at a future time.

The ENFP analyst would focus on "the big picture," getting a clear appreciation of the difference in the organizational impact of the proposed versus the current systems. She would appreciate current system problems such as low user morale or inability to produce different views of the same information for executive meetings. Furthermore, a capable ENFP would be able to convince the marketing manager and others in the user community that she actually hears, acknowledges, and cares for their legitimate concerns.

It is also likely that an ENFP type working in the system development area would have a reasonably well-developed thinking (tertiary) function. We recall that the F in ENFP indicates a preference; lack of preference for thinking does not necessarily imply lack of thinking competence. This assertion can be made based on the fact that this ENFP analyst has chosen to work in the systems field, which obviously requires thinking. However, this person may not use the inferior function, sensing, often. She may miss consideration of details that may have an impact on the amount of time it would take or the resources that would be necessary to develop the proposed system.

An adequate appreciation of Myers-Briggs types and an understanding of how different dimensions can affect system development would contribute significantly to a *functional synergy* between the ISTJ and ENFP analysts. Types need other types, and this truth becomes even clearer when considering order of preference. When considering the broad spectrum of tasks, both structured and attitudinal, that are required for an adequate requirements determination, the ISTJ and ENFP can consciously cooperate. Each can be aware of his/her strengths and areas where he/she needs assistance. They can be open in discussing these with each other. They can learn from each other during this development phase, emphasizing their strengths, strengthening weaker points, yet respecting the boundaries of how far they and their partner can "stretch" beyond their preferences. This type of awareness, coupled with a desire to cooperate at a deeper level, is a prime example of "emotional intelligence" at work in systems development profession. However, to operate most effectively, such awareness should be coupled with a degree of vulnerability, humility, and necessary trust, all factors related to deeper personal growth, a topic discussed later in this book.

The order of preference can also help to explain the most frequently and least frequently found types in the IT profession, according to earlier mentioned surveys. Common types found in the Lyons, Australian, and my own surveys are ISTJ, ESTJ, INTJ, ENTJ, ISTP, and INTP. If we list the top two preferences for each type (see Table 8), we see that the thinking function is prominent among all these types, along with either intuition or sensing. By contrast, in three of the four least frequent types among IT professionals, the feeling function is predominant, and in one of the four it is secondary. It seems evident that "high feelers" have been more attracted to occupations that allow for a more central role for the feeling function than does system development.

Table 8: Preferences of the Six Most Frequent and Four Least Frequent Types in IS

Type	First Preference	Second Preference
ISTJ	Sensing	Thinking
ESTJ	Thinking	Sensing
INTJ	Intuition	Thinking
ENTJ	Thinking	Intuition
ISTP	Thinking	Sensing
INTP	Thinking	Intuition
ISFP	Feeling	Sensing
INFP	Feeling	Intuition
ENFJ	Feeling	Intuition
ESFP	Sensing	Feeling

Preferences by Quadrants

Having addressed order of preference, it is useful to consider four quadrants of the MBTI Type Table: IS (upper left), IN (upper right), ES (lower left), EN (lower right). The four types in each quadrant have the same choice in the first two dimensions, and are thus somewhat related. We now consider how a member of each quadrant is likely to work.

The IS (quadrant member) focuses on the practical details. The focus in the organization is largely on the continuation of the status quo, on stability, on the "tried and true." This is largely because intuition is either fourth or third in the order of preferences for each type in the quadrant. In systems work, these are the loyal, trustworthy workers with particular ability to attend to detail. However, these would not usually be the types to promote assertively the adoption of new paradigms, methodologies, or technologies.

The ES quadrant also has a practical focus, but instead of practical *considerations*, here we see a tendency for practical *action*. People here are mostly focused on tangible results and offer leadership by doing. In systems work, people in this quadrant may excel at lower-level project management, where they focus on individual tasks and motivate to have results on time.

These are the "doers" among systems people, and the doing, for them, involves considerable interaction with others. Although such people thrive on action, they do not necessarily thrive on change. For all four types in this quadrant, sensing (with a sense for the "here and now") is the dominant or auxiliary function. Intuition (capacity for new possibilities), as for the previous quadrant, is third or last (inferior).

The IN quadrant consists of people with vision, who contribute *ideas*. Such people are valuable in systems work in system planning, higher-level modeling, problem/opportunity finding, and providing a solid basis for change. These are the "thoughtful innovators" who have shown a significant representation in Lyons' study, but were not as prevalent in the other two survey samples discussed earlier. Types in this quadrant, particularly the thinking intuitives, are valuable in research and development activity, which is indeed sought after to support innovative forms of business activity. For these types, intuition is either dominant or auxiliary.

The EN quadrant contains less than 20% of systems professionals in all three surveys. People here have a focus for *carrying out change*. They often lead through vision and enthusiasm as well as action. These types may be more at home in the management aspects of IS rather than the "nitty gritty" work. They are particularly useful in orchestrating the planning and development of new systems, organizational re-engineering, and promoting vision for a greater impact of IS on the business.

According to Hirsh (1991), types in each quadrant have a different organizational focus: IS—continuity; ES—results; IN—vision; EN—change. This focus can be emphasized with a key phrase for each quadrant: IS—"let's keep it"; ES—"let's do it"; IN—"let's think about it differently"; EN—"let's change it."

A challenge to IS management is to foster a functional, creative tension among people from the different quadrants. Here, *type literacy can go a long way*. Types who are seemingly opposite with respect to organizational focus (such as IS and EN) could provide valuable insight to each other, in search of an optimal solution. For example, the decision to develop software in-house versus buying a package and customizing can benefit from both a continuity and change perspective. However, awareness of one's own strengths as well as biases, in addition to a healthy respect for the legitimacy of an opposing view, are necessary for creative synergy rather than defensive conflict. Simple *awareness* of type differences is a starting point. An understanding of the orders of preference from among the four functions appears to provide more convincing reasons for the legitimacy of varying views. Having participated in specific exercises intended to teach the effect of type differences in IS work would, of course, have ultimate impact on the motivation and capacity for synergy and cohesion.

FOUR TEMPERAMENTS

Psychologist David Keirsey (1978) identified four *temperaments* that can be derived from the 16 MBTI types: Promethean (Rational)—NT; Apollonian (Idealist)—NF; Epimethean (Guardian)—SJ; and Dionysian (Artisan)—SP. Thus, each of the 16 types belongs to one and only one temperament. There are specific life attitudes particular to each temperament and, of course, these attitudes are carried into the work environment.

The Guardian comprises about 38% of the U.S. population. A person of this type longs for duty and exists primarily *to be useful* to society. The SJ must belong, and this belonging has to be earned. He has a belief in and a desire for hierarchy. The SJ is "the foundation, cornerstone, flywheel, and stabilizer of society." This is the conservative traditionalist. As we have seen earlier, SJs make up the largest fraction of IS professionals.

The Rational is found in about 12% of the U.S. population. The NT values competence and loves intelligence. He wants to be able to understand, control, predict, and explore realities. He often seeks to study the sciences, mathematics, and engineering—what is complicated and exacting. NTs tend to live in their work and to focus on the future, having a gift for the abstract. They have the capacity to think strategically and to develop visions of the future. They work on ideas with ingenuity and logic. They can be self-critical, perfectionistic, and can become tense and compulsive when under too much stress.

The Artisan is found in about 38% of the U.S. population. She is impulsive, living for the moment, wishing to be free, not tied down or confined. She has a hunger for action in the here and now. SPs are spontaneous, optimistic, and cheerful. They thrive on variety, and can be easily bored with the status quo. They also have a remarkable ability to survive setbacks.

The Idealist is found in about 12% of the U.S. population. She is the deepest feeling person of all types and values deep meaning in life. The main need of this type is authenticity to one's deepest self. NFs speak and write fluently, often with poetic flair. They seek interaction and relationships. They enjoy bringing out the best in other people. NFs work towards a vision of perfection and can be unreasonably demanding on themselves and others.

In certain situations, 16 MBTI types may be considered too many. The four derived temperaments, each consisting of four MBTI types, thus offer a more aggregated approach which nonetheless highlights key differences among people. Thus, the Kersey temperaments can make a valuable contribution to the systems development area.

Temperaments in IS Work

Table 9 shows temperament frequencies in the three surveys mentioned earlier. In two of the surveys, the SJ-Guardian temperament is overwhelmingly the most frequent, with the NT-Rational in second place. In the other survey (by Lyons), the NT slightly outnumbers the SJ. In all three surveys, the frequency of SPs and NFs together was under 22%.

As we consider the key strengths of each temperament: NT—vision, competence, abstraction; SJ—detail, practicality, facts, organization; SP—spontaneity, practicality; NF—authenticity, relatedness, we see readily that each of these strengths is very welcome in the systems profession. Information systems development comprises a variety of activities and tasks, and the strengths of the different temperaments are instrumental in these different tasks.

Once we consider the issue of the task-temperament match in more depth, we begin to appreciate that literacy in this area can indeed have an impact on the workers' energy and productivity. Detailed work in *attempting to link specific system development tasks with temperaments* have been done by Patricia Ferdinandi of Strategic Business Divisions in the U.S. Table 10 is reproduced from her insightful article (*Software Development*, 1994). Although the identified connections are hypothesized by the author, they are

Table 9: Temperaments in Three Surveys

	Lyons	Australia	Kaluzniacky
NT	40.7%	14.8	21.1%
NF	12.1	8.7	6.0
SJ	36.9	73.1	64.5
SP	10.3	3.3	7.5

Table 10: Matching Task to Type (from Ferdinandi)

Task or Function	Preference/Temperament
Identify the scope of the project	Intuitive, Thinking/Rational
Define the current process	Sensing/Guardian
Facilitation/JAD session leaders	Intuitive, Thinking/Idealist, Rational
Planning and architecture	Intuitive/Rational
Project-planning tracking	Sensing, Thinking, Judging/Guardian
Establishing business policy	Intuitive, Thinking, Judging/Rational
Essential business model	Intuitive/Rational, Idealist
Technology model	Sensing/Guardian, Artisan
Programming, testing, problem solving	Sensing, Perceiving/Artisan
Installation	Sensing/Artisan, Guardian
Organization and motivation	Feeling/Idealist
Network and communication design	Sensing/Artisan, Guardian
Entity Relationship model	Intuitive/Rational, Idealist
Logical data model	Intuitive/Idealist, Rational
Physical data model	Sensing/Guardian, Artisan
Data repository management	Sensing, Judging/Guardian
Brainstorming	Intuitive, Perceiving/Idealist, Rational
Structure diagram or charts	Sensing, Judging/Guardian
State transition diagrams	Sensing/Artisan, Guardian
User acceptance criteria	Sensing, Feeling/Guardian
Procedure manual	Sensing, Judging/Guardian
Training manual	Intuitive, Feeling/Idealist
Mini specifications	Sensing, Judging/Guardian
Scribe	Sensing, Judging/Guardian
Identifying objects or entities	Intuitive/Idealist, Rational
Identifying methods or processes	Intuitive/Rational, Idealist

made with an extensive awareness of both task and type. It is a call to MIS researchers to corroborate or challenge this information. We now try to extract, in an ordered fashion, the valuable insight from this detailed effort.

Rational-NT: The strength of the NTs is competence, abstraction, high-level vision. Their strengths are particularly desirable in systems planning, process redesign, and high-level modeling. NTs excel at applying logic to new possibilities, often in a creative manner. They can also abstract away and model essential elements of a business process and can identify, with thoroughness, inter-process connections. Thus, in Table 10, NTs are linked with tasks such as project scoping, system planning, establishing business policy, and data and process modeling. Such types are valuable in business process re-engineering. The introverted NTs will tend to think about business processes and future supporting systems differently from current consideration. Extraverted NTs will actually take the lead in carrying out change. They can be instrumental in planning and initiating systems projects at a high level. With the immediate future of information systems focusing on large- scale enterprise resource planning systems and on innovative Internet-based e-commerce applications, the contribution of NTs will be invaluable and requirements for such strengths will likely increase. To carry out such work without the NT may cause unnecessary risks. The SJ will lack the visionary, abstract orientation and could tend to focus excessively on details. The SP would likely find such long-term planning and modeling work as lacking spontaneity, action, and immediate realism. To the NF, extended hours of abstract logic without a human, relational focus might be draining and unsatisfying.

Guardian-SJ: The SJ temperament undoubtedly has a definite and likely permanent place in the IS profession: The Australian survey shows SJ types comprising 73% of the survey sample—three out of every four IS workers in the sample had the Guardian temperament. An SJ is characterized by detailed, linear thinking with an orientation to facts and the status quo. Order, structure, procedures, and belonging rank highly with an SJ. Thus SJs have traditionally been attracted to and have excelled in procedural computer programming, and have often been promoted from programming to other areas of systems work (analysis, design, project management, etc.). In Table 10, SJs are linked with lower-level modeling of the current processes (when the need for detail supercedes the need for abstract conceptualization of future possibilities), program structure charting, physical data modeling (where the real is emphasized over the conceptual network) and technology design, programming, testing, listing user acceptance criteria, creation of the system procedures manual and system installation. We can see that each of these tasks requires detailed, logical thought with practical consideration. Such tasks, therefore, are the natural domain of the SJ; from them, he derives energy and motivation.

An important question for the future of the IS profession is whether the sphere of influence of the SJ is shrinking. In days of basic, functional, transaction-oriented reporting systems mostly constructed in COBOL, the SJs predominated in the systems profession; many people of other temperaments may not have found such a profession attractive. However, the times have certainly changed. The systems area encompasses a much broader scope of activity and competencies. Systems are making use of a variety of media and information types. Their influence is expanding significantly and their role in many business organizations has become more central. Thus, while there remains a significant proportion of "SJ activities" in systems work, room for meaningful contribution from NTs and also SPs and NFs has increased. The profession has become much more diverse.

Artisan-SP: This temperament was found in 3-9% of the survey samples mentioned. Although not that many SPs appear to be in the IS profession, their contributions cannot be undervalued. They value practicality and spontaneity, and do not like to work at only one task at a time. Thus, SPs are ideally suited to maintenance ("firefighting") work, because of the variety and unpredictability. SPs can also be found at help desks. Also, SPs can be valuable working beside SJs on tasks requiring a practical, functional orientation. They may provide valuable balance to SJs because of their open-endedness. Thus, in Table 10, Artisans are listed alongside Guardians for tasks such as network design, physical data modeling, and particularly, installation, where unpredictable difficulties will be handled as energizing challenges. With the introduction of visual system development environments and multimedia systems, SPs may indeed be attracted in larger numbers to the systems field, once they can be assured that they will not be drained by routine and boredom.

Idealist-NF: NFs are also not numerous among systems professionals; 6-12% were reported in the surveys. An NF is a person of deep feelings which are either dominant or auxiliary and mostly ahead of thinking. NFs must be authentic to their deep self when they are working and have a strong relational orientation. Usually NFs in systems work, although they prefer feeling, are also quite capable in the thinking function. Thus, they can have a unique and significant role, if they are recognized for their particular talents. Many systems projects, to be successful, need effective human relations at several points: initiating the project and obtaining support from higher management; getting enthusiastic user support during requirements determination; enabling of cohesive teamwork during analysis, design, and development; and production of training materials that would be enthusiastically adopted by the users. NFs are adept at listening, persuading, motivating, and harmonizing. They are most emotionally insightful of people and very adept at perceiving non-verbal communication. The increasingly socio-technical nature of most systems efforts nearly mandates input from capable NFs. Such people are often sought after as managers and motivators of system professionals.

Thus, we see that each temperament is uniquely suited to specific tasks in the system development effort. All temperaments have something to contribute in IS. It is important, however, not to continuously try to fit the "square peg into the round hole."

PERSONALITY DIMENSIONS
AND SYSTEM DEVELOPMENT

In order to appreciate more directly the influence of personality characteristics on system development work, we now examine *key life cycle activities* and the strengths that MBTI dimensions can bring to such activities. This may clarify somewhat the perceived connections between type and task as presented by Ferdinandi in Table 10.

System planning can be carried out more formally or less so. A group of higher-level persons from organizational management and from the IT function will meet to discuss the strategic direction of the organization for the next few years. Then, IT support of each new area of organizational activity is examined. New system development efforts are identified at a high level and perhaps prioritized. Overall feasibility discussions, in particular budget requirements for new IT projects, may be addressed.

In such meetings it is essential that the IT representatives be competent and credible. They must also "speak the language" of management. Intuition is a key personality characteristic in such situations. It looks not so much at the factual, at "what is," but rather at "what might be." When discussing innovative IT support for new organizational initiatives, such possibility thinking is essential and this is the strength of the intuitive. The intuitive thinker (NT) excels at identifying technological possibilities and doing so with confidence. Based on current problems and new opportunities, he conceptualizes an idea for a new system and can assess its impact on the organization. The intuitive feeler (NF), who should also have a well-developed thinking capacity, has her own valuable contribution to make in system planning. Such a person is aware of the non-verbal communication of the management side. She is sensitive to the "sore points" of management and is genuinely interested in harmony during such meetings. She is able to persuade and reconcile by using the appropriate words with appropriate intonation, body language, etc. Such activity energizes her, and the energy as well as the substance of her communication is welcomed by the management side. Thus, she is adept at winning the trust and respect of management. The NF's contribution can go a long way in creating an atmosphere of partnership in strategic planning meetings. Whereas the NT is focused on the intangible realities in technology and strategy, the NF will focus on intangible realities in human communication and human attitudes. Thus, the NT and NF make up a truly socio-technical team.

A few words can also be said about the E/I and J/P dimensions in strategic system planning. The extravert thinks as he speaks and can usually generate ideas "on the spot" in a conversation. The introvert needs a time of solitude to prepare his ideas before actually communicating them. Strategic system planning is not usually done in one meeting. Thus both the E and the I have an opportunity to contribute from their strengths. However, it may be wise for IT to assess this dimension among the management representatives in the planning meetings. If higher management is represented by strong extraverts, an IT team consisting solely of introverts may be viewed as weak, unresponsive, or disinterested. Ideally, in such a situation, an INT paired with an ENF could provide a welcome balance on the IT team (this is another example of "emotional intelligence" —a term expanded upon in Chapter IV—within IT).

A J is more interested in structure and closure; a P prefers to weigh many options and remain open-ended. High-level planning meetings consist of many often open-ended ideas that may be created spontaneously; definite conclusions may not be reached for some time. Thus, a P may seem more suited to such an environment. However, a P (e.g., an NTP) may be creative but indecisive. He may propose many possibilities, but may have difficulty recommending one of them without extensive consideration. If the P is aware of this possible weakness, however, he may indeed be best equipped to provide impactful IT representation. A J, on the other hand (in this context, likely an NTJ or NFJ), may struggle if a meeting is not "going anywhere," and may try to force planning efforts to premature conclusions.

After the planning effort identifies the need for a particular information system, an important step is *requirements determination.* This is an analysis of *what* is needed in the new system: chiefly what data, what processing, and what user interfaces, both for input and output. Such an analysis is usually done in levels, especially for an extensive system.

Often, existing business processes may need to be redesigned (if not completely re-engineered) or new processes added. The substance and sequencing of the business processes to be supported by the new system must be modeled. One can start with a high-level business model and decompose this into more basic functions and then into processes that these functions comprise. When considering new or redesigned organizational functions, we have a definite opportunity for possibility thinking. However, such thinking now contains an element of structure, and this structure becomes more prevalent the lower down we proceed in process description.

In *modeling*, intuitive thinking (NT) is very useful. The NT can conceptualize new ways of carrying out business activity to reduce inefficiency and can represent this in structured, diagrammatic form. However, modeling does imply some degree of closure. At a high level, where one depicts basic business functions and their interrelationships, there is structure but yet considerable flexibility. Not that much is yet "pinned down." Yet, this structure needs to be thoroughly thought-out and fairly complete, for on it will rest more detailed modeling efforts. Also, this structure is based on possibilities for future functioning which differ from current realities. Such high-level business modeling is firstly the domain of the NTP. While considerable time is spent alone, there is a definite requirement for interaction with management to verify the correctness of the models. While, in some cases, some of the modeling activity can be carried out during a meeting with management (e.g., Joint Application Development sessions), often, detailed modeling is done on one's own and is then followed by a presentation to organizational management at a specific, arranged time. Thus, both extraverts and introverts can be involved here.

Once there is a high-level business activity model and a corresponding model for the supporting system's functions, more lower-level modeling will be done for the new system. Here, there is considerable need to define precisely new processes and to decompose them to even lower levels. This activity involves both conceptualization and structure, but with increased emphasis on structure. This is a likely domain for the NTJ (quite possibly an INTJ). There is considerable substance and an increased level of detail in such models. The INTJ is capable of single-minded concentration, has superior analytical ability, and needs to see concepts worked out and applied. Such a person's thirst for judging and closure allow for ideas generated at a higher level to be processed towards a specific goal.

It must not be forgotten that modeling the functions of a new system beyond the high conceptualization level involves understanding current business and system activity to an increasing level of detail. While this is not the natural domain of the NTP, it can fit the NTJ to a greater degree. The expert, however, in analyzing current procedures is the STJ (possibly an ISTJ). Here, there is not the emphasis on creating new processes, but on understanding (and documenting) thoroughly what is going on now and where the problems or inadequacies exist. There is, therefore, an existing background structure for this work, something that an STJ needs. The detail-orientation of the ISTJ in particular is a major strength in representing adequately what is now going on. A model for an improved system usually builds on the basis for the current system. There is an opportunity for effective STJ-NTJ cooperation. If each person was explicitly aware of his order of preference (Table 7) and that of his partner, and both could recognize this order operating in their daily work, both would indeed develop a new level of "emotional intelligence" in the IS field.

Model building for a new system requires considerable interaction with various organizational personnel. The modeling team needs to understand current processes and the limitation of current information. It needs to appreciate the feasibility of new system possibilities and needs to verify its models with middle and higher management. Here, again, the intuitive feeler (NF)—possibly an ENF—has an influential role. The NTP, while excelling in vision and abstraction, may be insensitive to people's feelings or to significant political realities. The NTJ may be driven single-mindedly to see his concepts applied, and the STJ may be so focused on process detail so as to disregard human, relational aspects of modeling work.

For example, when ascertaining how a credit application process works in a company, after speaking to the credit manager, an analyst might say:

"So, you're telling me that after a customer files an application...." On the surface, the statement seems truthful and innocent. However, the manager may feel she is being put on the spot and that any mistake in what the analyst is recording is her responsibility. An NT or an SJ orientation may not readily see this, but an NF likely would. Instead, the NF might naturally remark: *"As I understand it, after a customer files an application...; have I understood this correctly, if not please help me out."* A seemingly minor point, yet the tone of communication "colors" a relationship between the system developers and management/users. The nature of this relationship can have a significant bearing on the success of the project. It should be noted, however, that such effective communication, while natural to NFs, need not be their sole province. Other types, such as NTs and SJs, should also be able to develop such competencies in an attempt to work with emotional literacy and "professional wholeness."

Apart from modeling processes, data for a new system also needs to be modeled. The organization may already have a "corporate data model" that applies to all existing systems. In such a case, only alterations or additions to this model need to be addressed and modeled for the new system. Alternatively, the new system may have its own complete data model, since its database may be independent of other systems. In many cases the data model for the new system is not radically different from that for the old system; it is the processes that change more in the new system.

In *data modeling,* it is clear to see that thinking (and likely judging (structure)) is of considerable help. Usually, conceptual data modeling is carried out first, possibly using the entity-relationship approach. Here, intuition is useful, as entities must be identified from the domain of application and attributes must be assigned. Such work can be quite abstract—there is not always a defined procedure for it. Two different analysts might present somewhat differing but adequate models for the same situation. It is not immediately clear whether there would be a significant advantage in a perceiving or judging approach here. The model of a perceiver may be more elegant since he is not impatient for closure and may consider a variety of possibilities.

The logical data model maps the entity-relationship diagram to a conceptual database schema. For a relational database design, such a schema would indicate what fields would be contained in which tables, and where the key fields and the common columns between tables would be. Here the goal is to design the database for efficient storage and retrieval. Considerations here are quite structured and a J preference may find this work easier; an NTJ or an STJ could find such work rewarding. The physical data model involves considering how the database management system will actually (physically) implement the schema. These are detailed practical considerations, likely the domain of an STJ.

Design of the *menus, input screens,* and *outputs* would likely appeal to a sensing personality, an SJ or possibly an SP (who might enjoy the variety in the design of graphical user interfaces). Design of *program logic* would likely attract an STJ who is likely also introverted; an INTJ may also find this work acceptable because of the structure and the potential for creating original, insightful algorithms.

The actual *programming* has traditionally attracted ISTJs, and to some degree ESTJs. ISTPs may tolerate programming, especially if it is within the context of a visual development environment. There is considerable linear thinking required in programming. An ISTJ seems ideally suited to working alone, thinking factually and step-by-step with considerable structure. The sensing component does not thirst for constant novelty and does not mind programming similar concepts with a similar flow of logic on many successive occasions. It should be noted that an NT would certainly have the capacity to do programming, but may require a more challenging and changing environment; otherwise, the work may become uninteresting for such a type.

Program testing involves both structural and functional testing. The former builds test cases based on knowledge of the program structure, to ensure all statements and conditions are tested at least once. Structure, practicality, and thoughtfulness are evident here—all in the ISTJ domain. Functional tests derive test data based on what functions the program is supposed to carry out. Here, practicality is paramount, and there may be more variety here as well as some contact with users. Hence E/ISTP and also E/ISTJ would find testing rewarding.

Since a considerable number of systems today are being built using some form of a *"rapid development"* approach, it is worth considering this phenomenon from a personality point of view. Instead of progressing through the development stages of analysis, design, and system development (coding) in succession, visual development environments (such as Visual Basic or PowerBuilder) allow the analyst to develop menus, screens, reports, and even data files quickly, largely dragging and dropping controls such as labels, text boxes, and buttons on a screen surface. Such controls can then be easily moved, deleted, or changed. The environment itself generates much of the code for these system components.

This capability allows for the analyst to consult with the user regarding the requirements for a subsystem, and then to design and construct, possibly with the user's assistance, a basic prototype of the subsystem with the desired menus, screens, queries, reports, etc. Then the stages of analysis, design, and development are repeated on the prototype subsystem several times until the final version emerges.

From a psychological point of view, such rapid development activity provides user interaction, tangibility, some spontaneity, and variety. We can automatically associate these with extraversion, sensing, and perceiving; thinking and feeling appear more balanced here. If we consider alternative media such as text or video, this may confirm the interest of types such as an ESTP. We are, therefore, motivated to ask whether visual, interactive development can indeed attract types other than the traditional ones (STJs, some NTs) to the IT profession in significant numbers.

System *installation* (hardware and software) is essentially practical activity. At this point, innovation is not central. Also, the installation and operation of the new system requires structure and precision. These characteristics point to SJ. However, considerable activity may need to be carried out in parallel, and one person may need to oversee several tasks at the same time. Here, then, is a scenario for the SP.

Training the users on the new system benefits considerably from a developed feeling function. The focus must be more on the user than on the application system. Here, feeling types have a significant contribution to make. While the production of training materials (manuals, CDs, etc.) may benefit from the intuition of an NF, the actual training activity could be as well carried out by an SF type with a good capacity for extraversion.

Maintenance activity is often compared to "fighting fires." In addition to planned changes or enhancements to the system, many maintenance requests arise as a result of some system malfunction. Dealing with unpredictability—practically, in a thinking manner—is the forte of an STP type. Such persons may actually feel stressed or bored by doing only one task at a time. In the maintenance area, STPs have the opportunity to attend to several tasks simultaneously.

Observed IS Work Preferences and Type Dimensions

Following are descriptions of work preferences that I have observed within a local sample of four IS professionals. This group participated in the creation of an amateur video on the use of MBTI in systems development work.

An introvert needed to go away and "digest" material alone after talking to a user. An extravert liked working in groups better than working individually; this person also disliked using e-mail as he found it to be an impersonal form of communication. The introvert, of course, did not mind using e-mail.

For a sensing person, generating new alternatives, without having a specific reference as to how others have done it, did not come easy. She would rather use the ideas of others than generate them. However, she could pride herself in being good at grasping facts and detail.

An intuitive analyst was emphatic about his need and liking for creativity and innovation, but not liking detail. Also, the sensing and the intuitive regarded the task of procedural coding differently: the sensor viewed programming as factual, whereas the intuitive experienced it as conceptual.

The thinking type was very much attuned to the logic in a system, whereas the feeling analyst explicitly stated he was uncomfortable with people that were "overly logical." He liked satisfying users and was careful to consider "where people were coming from" when making a statement.

The above section addressed the main system development activities from a personality/temperament viewpoint. The discussion was somewhat heuristic and idealistic at the same time. More research is required to corroborate the various suppositions. At the same time, how practical is it to follow the above prescriptions? The material was presented firstly for orientation and consideration, particularly by project managers. While such specific matching of tasks and personality traits may not always be feasible in practice, at least it can be realized where specific strengths are advised and where a significant absence of these strengths may cause difficulties.

MANAGING IS PERSONALITIES AND TEMPERAMENTS

Apart from a greater self-awareness or the part of individual programmers, analysts, etc., a significant opportunity to actually apply MBTI in IS development is presented to the *project manager*. The above sections provided considerable insight as to relating type to specific system development activities and tasks. We now examine more general approaches to managing the four temperaments and refer to frequent IS types with these groups.

SJ: An SJ requires a stable and orderly work environment and relies considerably on standard operating procedures. He is attentive to detail and sensitive to meeting deadlines. A judging analyst (also sensing) would take meticulous notes and focus on detail. She wanted to see completion in her work, preferred to work on one thing at a time, and was very frustrated when an interview wasn't going anywhere.

However, a perceiving analyst did not appreciate people "with one right answer" and was actually stressed by "too much predictability." She needed to do analysis work in addition to programming, since the latter was too structured. Also, the perceiving analyst did not need to see immediate practicality in what she learned.

An interesting and insightful story can be related regarding the last point. A student of ESTJ type had enrolled in an undergraduate computer science program. For one of the assignments, she was to write a procedural (3GL) program to produce a "stick man" who would dance on the screen. Outraged at such an assignment that she would never use and a parallel teaching philosophy in general, she quit the program and registered in a smaller university in an applied program in business computing. On the contrary, an ENTP analyst stated that he would find the assignment to generate a dancing stick man quite interesting and challenging.

Such practical insights do tell a story. They corroborate, for the inquiring mind, a practical effect of type on attitudes and preferences in IS work. Moreover, these observed attitudes as discussed above are quite consistent with MBTI type theory. The SJ person is loyal and trustworthy, and appreciates being trusted by the manager with responsibility; the feeling SJ types appreciate frequent praise, whereas thinking SJs primarily need to be treated fairly. They do not appreciate the manager's changing his mind frequently and without detailed justification. In general, SJs do not appreciate surprises. In dealing with others, SJs find it difficult to respect people who do not follow schedules or prescribed procedures.

Two common IT types from the SJ group are the ISTJ and ESTJ. The ISTJ is task-oriented, organized, practical, careful, determined (Isachsen & Berens, 1988). The ISTJ needs organization and a structure in order to work. He does not tolerate ambiguity well. He needs to understand his task, which should be structured and detail-oriented with a goal of preservation and verification. ISTJs may not like to work on several tasks simultaneously, as closure is very important to them.

The ISTJ can be counted on to follow all prescriptions of a methodology and to preserve the established approaches. When asking this type to comment on an idea or document, be sure to give him enough time alone to prepare a thorough response. When considering a major change, the ISTJ can provide valuable input as to possibilities for failure that will need to be addressed.

When managing an ISTJ, tasks, times, and desired results must be communicated clearly. Give them opportunities to organize and structure, in detail. Treat them fairly and predictably, without introducing new requirements abruptly and unexpectedly. Be sure they know what the desired result of their work is and how their work will be assessed. When trying to develop a culture of "emotional literacy" and "soft skills," this type will not be ready to adopt these easily since this area is not as tangible or structured. Backing up claims of impact of soft skills with measurable, credible data would be very desirable.

On the other side, an ISTJ as a manager is loyal to the organization and manages predictably and somewhat impersonally with an attempt at fairness. He will likely be task-oriented rather than person-oriented in approach. He will run efficient, planned meetings and try to adhere to an agenda, being upset with people who go off "on tangents" and speak spontaneously. He is very thorough in his area of expertise and is willing to accept considerable responsibility. He is also sensitive to omissions, discrepancies, and deviation from the plan. The ISTJ manager may, however, ignore the needs and sensitivities of feeling types, particularly NFs. Also, he may be excessively upset with visionary NTs and spontaneous SPs in their ignoring of or being ambivalent to standard operating procedures.

In IS such a person might make a better lower-level supervisor than a high-level manager, especially without explicit management training. As one progresses higher in the hierarchy, issues become more intangible, ambiguous, and unstructured. Strategic planning or persuading a user manager to abandon a subsystem's development, for example, require more development than factual, structured logic. It is at this level that IS managers have received considerable criticism in the literature. This is quite possibly because of such a high incidence of STJs in IS who have progressed to senior managerial ranks.

It is thus not difficult to see that in order to be an effective IS manager, the ISTJ type needs to grow, to acquire a broader perspective as well as "emotional intelligence." But for such a type, the *feeling* needed in such a growth is the tertiary function behind sensing and thinking, and may therefore not be used often. Such a type's emphasis on traditionalism and the status quo, however, may strongly discourage a serious interest in personal, psychological development (until, perhaps, such development becomes part of the "status quo").

The other SJ type found commonly in IT is ESTJ. This type is also practical, structured, and thinking oriented, but is characterized more as a "doer." She is decisive, organized. and wants "to get things done" with no time wasted. Such a type requires more interaction with people and thus has a natural advantage in teamwork and user interaction. Also, the ESTJ needs a precise definition as to what needs to be done and how, before committing her natural talents to getting the job done.

The ESTJ's extraversion along with task orientation are advantages in being a manager. However, this type may, again, experience difficulties because of being too "trigger happy" and possibly narrow-minded, impatient, and insensitive. Consequently, growth and a deeper appreciation of psychological factors in IS work are, again, in order. However, this is not likely to be a natural priority for the ESTJ manager. While the ISTJ may hesitate to do growth work because it is not traditional enough, the ESTJ may balk because it is not practical enough with immediate, measurable results.

SP: Like the SJ, the SP is practical and focuses on tangible issues. However, the SP requires spontaneity and independence at work. She appreciates going through the process more than arriving at the end result of the process. An SP likes "to do her own thing," but with a practical, tangible slant. Such employees like variety and may learn new software packages just for the experience. With an enlarged scope of IS influence and an expansion of information technologies, the SP orientation is indeed valuable to the profession.

For the SPs, troubleshooting is an adventure and help desk work would appear natural. Maintenance is another area of strength. The SP actually prefers to be working on several projects in parallel and is excellent in crisis situations. She is willing to dive in and try new approaches. Instead of being overly concerned with schedules, she concentrates all her efforts on whatever she is doing at the moment. Consequently, the SP type is not bound to standard operating procedures. Being managed in a very controlling way will cramp the SP's style and drain her energy. However, the freedom to work independently will cause this type to flourish. SPs will not appreciate those who are excessively rigid regarding established procedures or those who philosophize without showing results.

Of the SPs, it is more the thinking types that are attracted to IS, and more of them would be introverted. The ISTP is probably more suited to maintenance and the ESTP to the help desk. ISTPs, in general, enjoy learning new tools and will do so more eagerly than the conservative ISTJs. They are oriented to the task in the present and can thus be oblivious to emotional or intuitive issues. Their adaptivity, practicality, and introverted focusing are definitely talents essential within IS and should be recognized as such by management.

As with STJs, where the ISTP requires a variety of tasks with considerable freedom, the ESTP in addition values action, particularly one that creates success and makes visible impact. He is very resourceful and can come up with ingenious ways to get things done. ESTPs' extraverted nature often leads them to display excitement and optimism, allowing co-workers to feel they are in a "win-win" situation. One can imagine the value of an ESTP project manager in a typical over-time, over-budget systems project.

The STP as a manager will differ from the STJ chiefly through *flexibility*. He will not "worship" standard procedures for their own sake, modifying them as necessary. He will handle crises more naturally than the STJ and may be more adept at negotiating, with his capacity for "give and take."

The STPs would not find personal development threatening because it is not traditional, but they may question whether it is really practical and impactful. To engage STPs, any such growth efforts will need to address this issue.

NT: The NT is the visionary, the strategist, the possibility thinker. NTs value competence and look for competence rather than title in a manager. Contributions of an NT come from thinking change and promoting change. Information systems are built to incorporate changes and often enable significant changes in organizational procedures. What would such an IS environment be without an NT?

An NT needs to be trusted with competence, and his often-ingenious developments need to be assessed fairly with in-depth understanding. An NT needs room to brainstorm and to create. To maintain the status quo (as with the SJ) is to stagnate, and to be practically flexible (as with the SP) is to miss out on the real meaning that lies in new concepts and new procedures. An NT will not appreciate routine, detailed, day-to-day activity nor standard operating procedures. He will not like being praised for ordinary tasks. He needs considerable freedom. His commitment level is often very high, but it may be more directed towards his ideas than towards the organization (particularly for an introvert).

There is a variety of NTs present in IT. The INTJ, an engineer type, is adept at model building and needs to see his work translated into action. As stated earlier, he is particularly suited to process and data modeling. He often feels compelled to implement changes that lead to more efficiency. He can apply enormous willpower, but can be single-minded and insensitive. The INTP is a more theoretical type, often found in research. He is skeptical and may often question authority. The current IT climate is very fitting for INTPs who can be managed with appropriate understanding. A drawback of such a type in an area other than pure research is, however, that his work may be too elaborate and perhaps too abstract to be of immediate use in an operational environment. The INTP needs an opportunity to learn continually.

The ENTJ possesses a combination of skills in analysis, goal-orientation, possibility thinking, and communication. Because of the extraversion, she may be more interested in working for the organization rather than just for the sake of ideas themselves. However, the purpose of communicating with people will not have a primarily people-focus, but a change-and-action focus. An ENTJ must be appreciated for suggesting change and should be given appropriate opportunities to lead. The ENTJ is a relatively common executive type who can determine and communicate how an organization can change.

The ENTP's main difference from the ENTJ, of course, is a desire to remain "open-ended." This type is well suited to strategic planning and brainstorming in groups. He would like to make key visionary contributions in a project and then move on to another, leaving details to others.

The ENTP, by synergizing possibility thinking with open-ended flexibility, can make major contributions in alignment of systems strategy with business strategy, as would be required, for example in the implementation of Enterprise Resource Planning systems. Such a type should not be assigned routine tasks or even variety without the visionary, conceptual component.

As managers, NTs worship reason and focus on concepts. From here, there is some variation among the NT types. The ENTJ is an action-oriented visionary who can take charge with assurance and generate others' confidence. He can, for example, guide a complete reorganization of the corporate IS department. The INTJ can be more autonomous, making decisions carefully (often fully utilizing analysis of alternatives from Decision Support Systems). He may enjoy small teams of highly competent collaborators. His single-mindedness can sometimes undervalue consensus.

An INTP manager may communicate the general outline of her vision and allow considerable autonomy beyond that. In this approach, she may not provide enough details to a subordinate who requires them. If she perceives impending failure on the part of subordinates, she may take on the work herself, quite possibly disregarding standard operating procedures. In IS, the INTP may not aspire to a management position, preferring the research avenue.

The ENTP is more oriented towards organizational dynamics. He can competently keep several issues in his mind at one time, and prefers to be in charge of innovative projects rather than routine management. Such a type is very valuable in IS consulting at a higher level.

Since NTs are oriented towards abstract thinking, their weaknesses as managers lie in a lack of focus on detail, deadlines, or on interpersonal dynamics. They may ignore completely, for example, the need for praise and support among introverted feeler types. NTs could be encouraged toward personal growth, but they would first need to dissect all the principles of such growth with a skeptical, analytic approach.

NF: The NFs, while not exceptionally numerous among IS professionals, are "a breed unto themselves," and thus warrant particular management competence. This is generally the type with the deepest feelings, which operate "front and center" with a person. Thus, the prime need of an NF is to be authentic, that is, not to ignore or repress deep impulses and personal convictions. An NF wants to be first a person and then a professional. The

NF would like to be valued personally, praised (justifiably) frequently, and heard deeply both in terms of feelings and ideas. NFs like to see management take a deep interest in the welfare of the employees, both through personal interaction and proactive policies such as sponsorship for courses and conferences. They like to feel harmony in a functional working environment. Personal or organizational dysfunction, if ongoing, can demoralize them and impact significantly on their effectiveness.

The NFs have a significant people orientation and would like this appreciated. They are sensitive to moods on a project team or to hesitancy in the user community. They can excel in user interaction and in production of effective training materials for users. To violate them would be to treat them as nameless, faceless robots capable of churning out creation of pure logic. They are frustrated and even hurt by impersonal attitudes.

The NFJs are more likely to be found in IS than NFPs. ENFJs can contribute people skills more with a variety of users in the planning, analysis, and training stages. INFJs may be more content to apply their people skills in the context of a development team.

As for personal growth, one need not coerce NFs! They are unquestionably the leaders in this orientation and will be very pleased if their employer "finally" promotes psychological literacy at work. It is only more recently that the significant and unique contributions of NFs are beginning to be recognized in the information system area, which is now acknowledged as being socio-technical in nature.

Effective IT management does require attention to individual differences, more formally or informally. The IT manager, often with only a technical training, can no longer ignore the particular needs of his subordinate. Also, he can no longer ignore significant, detrimental biases in his management style, which may be largely personality based.

Having digested only some of the preceding material, we can envision a variety of situations, e.g., an NT manager and an SJ subordinate or vice versa, a steering committee of all SJs, or an SP doing only C programming. We can appreciate where, in such situations, significant potential energy is being lost, largely due to ignorance. Whatever we may feel about the rigor (or lack thereof) in type consideration in IS management, most of us would agree that such considerations are indeed warranted.

RESEARCH AND
APPLICATION OF MBTI IN IT

The Myers-Briggs Type Indicator has become very popularized in recent years throughout the world and across many disciplines of study and work. And indeed it has begun to make its mark in the Information Systems field, both in academic research studies and in industrial application. We now examine highlights from such applications.

Noteworthy Research

We have already become somewhat familiar with the study by Lyons. He reports on a survey of personalities and work preferences of 1,229 computer professionals employed by over 100 different companies. Major tabulations have been presented in Tables 1 and 2. He points out about MBTI: "As with most programming languages, appreciation of its capabilities and usefulness increases with knowledge and experience."

Firstly, Lyons introduces the four MB dimensions and presents observed frequencies for each preference. He then segregates data by gender, showing more feeling types among women. He discusses sensing feelers and intuitive feelers as well as sensing thinkers and intuitive thinkers in the context of IS work.

Other valuable points:

- R&D organizations and firms that do a lot of state-of-the-art development attract and hire more Ns than Ss.
- A great many of the communication difficulties experienced on the job are based on the S-N difference.
- A difference in J-P attitude is second only to an S-N difference in causing communication problems.
- Feeling is an especially appropriate judging process when dealing with people, and it can be very helpful in supervisory and management positions.
- It is almost always good to have some diversity on the team in terms of psychological types.

Lyons ends with the statement that:

"simply being aware that a person may be of a different type and taking that potential difference into account can greatly improve communication" *[bold added]*.

A second research effort which had been mentioned earlier is that of Thomsett (1990). His focus is more directly on effective IS project teams. He reports figures for 656 IS professionals in Australia, where 63% of them are of types STJ. More detailed MB-related statistics have been presented in Table 6 and Table 9. His research team had applied MBTI along with two other instruments, the Job Diagnostic Model and the Belbin Team Role Model, in an IS team context within an organization with over 200 computer and related specialists. *Immediate productivity increases of 200% have been reported by the senior management of the computing group.* Such results cannot be ignored!

Basic personality statistics on system analysts from a large South African insurance company are presented by D.C. Smith (1989). The type table statistics are shown in Table 11. Here again, 64.8% of the analysts were of type STJ, a phenomenon of some concern, referred to by Edward Yourdon (1993) as a "cloning syndrome." Smith also notes that 81% of the types fall into a "conservative" category, and only 19% could be considered "innovative." Although the research is somewhat dated, such an occurrence now may indeed be a cause for concern, since e-commerce relies on innovative support from information technology.

Apart from mainly tabulative, exploratory research, Joy Teague of Deakin University in Australia (1998) has hypothesized which personality types would be best suited to each of analysis, design, and programming. She proposes NTs and NFs for analysis, NTs for early design stages, and SJs for latter design stages with ISTJs for programming.

She then presents statistics from a survey of 38 computer professionals who were MB typed and asked to rank their preference for analysis, design,

Table 11: Type Table Statistics (D.C. Smith, South Africa, 37 participants)

ISTJ		ISFJ		INFJ		INTJ	
	35.1%		8.1%		0		8.1%
ISTP		**ISFP**		**INFP**		**INTP**	
	2.7%		0		0		5.4%
ESTP		**ESFP**		**ENFP**		**ENTP**	
	2.7%		0		0		2.7%
ESTJ		**ESFJ**		**ENFJ**		**ENTJ**	
	29.7%		2.7%		0		2.7%

and programming. Proposed personality characteristics were then compared with observed results, and 85% of the people who preferred analysis were NTs or NFs. For design, 50% were NTs, 17% were SJs, and 17% SPs, with 17% NFs. Among programmers, 42% were SJs, 17% SPs, 17% NFs, and 25% NTs. Although the sample was rather small, the study did address an important question: *the issue of different types being suited to different tasks.*

The matching of type/temperament to task was proposed in most detail by Ferdinandi as described earlier (Table 10). This level of MBTI application seems the most useful, at least conceptually. Harnessing the proper energy for the appropriate task could have a significant effect on desired productivity. However, Ferdinandi does note that "any personality or temperament can successfully complete any task, but some types are naturally geared towards certain tasks and functions." She then considers the contribution of MB information specifically in the context of re-engineering business processes for which, then, a new information system will be developed.

Ferdinandi points out:

- Re-engineering is a conceptual effort dealing with possibilities; while this may be difficult for sensors to comprehend, it is ideal for intuitives.
- In meetings attended by intuitives, agendas will probably be served as a guideline rather than something to be followed strictly.
- Do not expect a detailed answer from an intuitive; gather the general concepts and put things into categories yourself.
- When reviewing details with intuitives, refer back to concepts and show how the details are important to the strategy.
- During re-engineering sessions, thinkers are particularly valuable since they are not affected by organizational boundaries or personalities (e.g., we can't even think of merging Joe and Bob's departments since the two don't get along well—such a consideration is not at the forefront of the thinker's consciousness, but it might cloud a feeler's proposal that two departments could merge in the future).
- Feelers serve as the team builders and understand how to keep people motivated during the long re-engineering effort. This is important, since the average re-engineering effort may last from three to five years before one can see noticeable improvement.
- Feelers may be better at selling the corporate strategies and new business policies developed by thinkers.

- When coming up with alternate solutions, the judgers, wanting closure, may "close the door" too soon; perceiving types will likely consider more possibilities and come up with more options for a given situation.
- While the SP artisans like spontaneity and change, they tend to look for improvements to the existing business processes instead of working painstakingly to re-engineer the entire business without immediate payback.
- The NF idealists are good at anticipating customer and user requirements rather than building a new system exactly to specification; NFs identify non-financial objectives that ultimately inspire vigorous efforts to improve the business process.

Ferdinandi also advises that when communicating with a person whose temperament preference does not fit naturally with the task at hand, an adept analyst will be able to use the communication style geared toward that temperament in order to be heard most fully. *Such a capability within the IS analyst is a direct example of application of "emotional literacy" within the IS area.*

More involved academic research involving MBTI and information systems had been carried out by Kathy Brittain White. In one study (1984), White compared two IS project teams that were given the same assignment involving the same user. One team was composed of all thinkers while the other contained 50% feelers. The all-thinker team did not produce a successful system. The system that this team produced did not meet the needs of the users. Also, communication with the users during system development was lacking. The developers spoke in overly technical language. With the 50% feeler team, the users expressed satisfaction both with the developed system and with the development process itself. They felt that this team was indeed concerned with their needs. Such an experience is consistent with type theory.

In another study (1984), White's results indicated that the MB types of team members and task structure both impact team effectiveness. White found that heterogeneity of types is best for solving unstructured tasks, but such diversity of types could be counter-productive in solving structured tasks. Her findings also indicated that one team might not be appropriate for all the stages of a project. In yet another study, White contends that "personnel awareness training can enhance change management and increase productivity."

Another noteworthy research effort is that of Kaiser and Bostrom (1982). The researchers note that personality characteristics of individuals involved

in systems development impact the way these developers perceive the organization, organizational members, and the function of information systems.
They also suggest that *system design reflects the design team's personality
styles*. In addition, they comment that feelers were often missing from teams
that were involved in project failure.

Noted IT consultant and educator Edward Youndon (1993) has also
referred to MBTI in IS. He quotes a seminar participant: "Simply *knowing*
the roles and the personality types makes everyone much more sensitive and
aware of the team's strengths and weaknesses." He then asserts: "The world-
class software organizations are, at the very least, *aware* of these team-role
and performance management issues (e.g., personality awareness). The aggressive ones are providing training to their professional staff in these areas,
and the very best have added professional experts—for example, industrial
psychologists—to their staff of consultants and advisors."

The above references to IT research involving MBTI are by no means
exhaustive. However, they do indicate a level of acceptability for this
personality system within the IT area. As well, the research has produced
noteworthy results which can have a significant influence on system development work.

Applications in Practice

Having examined research highlights, we now focus on some stories
"from the trenches," stories of actual MBTI application in IS practice.

MBTI has been widely accepted in the area of general management.
Thus, many organizations have been bringing in consultants for a half-day or
one-day workshop. In some cases, a follow-up workshop may be held some
time later. IS professionals have, in many cases, been included in such workshops and have thus been introduced to personality type concepts. However,
often there has been little follow-up to an initial workshop, and the typical
IS worker may have benefited only marginally from initial exposure.

Still, there are indeed situations in IS where MBTI was accepted more
seriously with impactful results. At Corning. Inc., MBTI has been used more
extensively, particularly in relation to teams. After initial introduction to MB
type, the IS staff go through four-hour training sessions that have participants
play roles, solve hypothetical problems, and listen to one another.

For example, they may gather all NFs in one group and SJs in another
and have the two groups solve the same system development problem.
Then each group shares with the other how it solved the problem, i.e., what
steps it took. These steps are then related to MB concepts. *This is a prime*

example of "emotional literacy" education in the IS area. Such heightened awareness has increased trust among Corning's team members, since they are aware of each other's perceptual differences and are not as likely to be intimidated by them.

Another significant application of MBTI occurred more recently at IBM, as part of their Team Pac program, used to train teams. MBTI was chosen as the personality component of the program because it had: i) acceptable validity studies and documentation, ii) books and conferences where people could learn more about type, iii) a self-scoring version for easy administration, and iv) considerable acceptance in the business world.

There were 26 topics selected for Team Pac and of those, three topics related well with MBTI—the three related to group work stages. Thus, IBM created three separate modules, each about four hours long.

The first module introduces type differences and gives participants exercises to understand each of the four dimensions. Then, the members respond to a series of teamwork dilemmas. This module ends with formation of a plan of action as to how this team will work with its differences.

The second module looks at the function pairs, ST, SF, NT, NF, since these are seen as influencing communication the most. People examine their communication styles and each person practices communicating with someone of another type. Then the participants are shown how famous business decisions were made and which of the function pairs were used in making them.

The third module examines leadership and learning styles based on the four Kersey temperaments. Team members' leadership preferences are related to the leader's temperament.

IBM finds that having three modules is effective, since people have time to absorb basic information and only then get into more depth: In just over three years, more than 30,000 booklets of the Team Pac materials for the MBTI modules have been used.

Such an extensive and focused application of MBTI is encouraging. The use of MB in the IT area may, at some point, need to be assessed through a maturity stages model. Many organizations have yet to introduce it; some have introduced it, but have left any follow-up to the individuals themselves; and a few, like IBM, have taken MBTI involvement to another level of maturity. It is likely, though, that unless type awareness is an ongoing effort in the IT environment, it will not realize its true potential.

Yet another noteworthy application has occurred at Hewlett-Packard in California. A team was created to implement SAP enterprise resource software. Considerable time was spent at the start examining communication and personality dynamics on the team. At first, team members were skeptical about investing such a significant amount of time on "touchy, feely stuff." But, as the group gelled, the people realized that they did not lose days, but actually gained weeks. *One HP manager commented that, for them, Myers-Briggs was a turning point!* Team members realized that "diversity and differences are what made the team successful in the first place."

Possible Uses

Having examined specific research results and industrial application efforts, one may have grown in the conviction that MBTI does have an ongoing, significant role to play in the careers of IT professionals. But, one can then ask where, specifically, might the Indicator be used and what tangible benefits can be observed.

The two most highlighted uses of MBTI in systems development relate to *teams* and *tasks*. In teamwork, it is important to be conscious of the reality that what energizes one person actually drains another, decreasing the person's effectiveness, productivity, and motivation. As a concrete example, two analysts at a finance company, one an ISTJ and the other an ISTP, were asked what stresses them most in the course of their daily work. The ISTJ replied that she was really stressed when she had a number of different projects to work on simultaneously. She would have much preferred to complete one item of work, put it away, and then start on another. The ISTP, on the other hand, indicated that she felt very much stressed when she was working on one thing too long. She actually needed to be working on several items at once and likely switching from one to the other spontaneously.

The above story is very significant, not only because it is true and agrees with type theory, but it is noteworthy above all because it shows *how the work energy of the different types, if not harnessed properly by psychologically aware IT management, will largely be wasted.*

In teamwork, therefore, assigning people tasks naturally suited for them is a primary issue. Following this is the need to balance teams such that each member's energy is utilized as functionally as possible. Such an orientation requires considerable awareness on the part of team members and managers, as well as deepened trust, in order for team members to be open enough to discuss genuinely their strengths, weaknesses, and occasions of drained energy.

Apart from task assignment, team formation, and team cooperation, MBTI can be used in individual career planning. Having developed considerable familiarity with one's MB type, and having observed and noted various ways in which one's type relates to one's work, an IS professional can look at her career in a more personalized way. Instead of adopting the "default" attitude of planning one's career strictly to maximize earnings and to achieve a prestigious recognition, an IS worker can examine areas which generate energy and enthusiasm, and furthermore, he can be able to explain why such areas make it possible for him to come alive. By one's taking such a view and having such a view acknowledged and actually encouraged by the employer's organization, a person will progress in his job to areas for which he appears naturally suited. In such an approach, higher earnings and more recognition, although not being primary motivators, are very likely to follow.

Type awareness can also be used for stress management and maintenance of wellness. Work stress often results from prolonged negative emotions resulting from work conditions being experienced as unfavorable. By understanding such emotions, an IS worker can explain his stressful experience in terms of type (and other areas of awareness such as those mentioned in following chapters) and have such an explanation respectfully received by type-aware management. Such communication could initiate whatever environmental changes are possible to alleviate individuals' stress.

A person could also experience negative emotion resulting from inner attitudes for which no immediate environmental adjustment is possible. For example, each of the four temperaments is identified with a basic need: competence (NT); authenticity (NF); spontaneity (SP); and organization, belonging, order (SJ). If such a need is frustrated significantly, the employee may experience stress. If no explicit organizational remedy is probable, the worker can, at least, recognize what it is that seems to be the main cause for stress and can attempt to make inner emotional adjustments based on increased awareness. For example, a worker who is constantly required to apply a software tool without having developed what she feels is sufficient competence with the tool, can redefine competence within herself. She can begin to see herself as competent in a relative sense, when she has appropriated at least the main features of the tool. The association of the stress with a feeling of lack of competence and the understanding of why competence is so highly valued would come from a deeper MB temperament literacy.

Another area of type application relates to end-users. IS professionals interact with users when determining system requirements and confirming their design (such interaction may become more ongoing during a prototyping exercise). Also end-users need to be adequately trained as the new system is put into use. Type literacy can be very useful here, although it may take on a slightly different character.

In determining user requirements, *it is essential to hear truly the user's message*—not to hear only the words, but *to grasp the intended meaning behind both the words and the non-verbal communication*. However, it is not likely that many systems analysts will have the opportunity to test and identify formally their users' personality type. Thus, type insight will need to be gained by observation and intuition.

An extraverted user may be defining system requirements as she speaks with the analyst, particularly if she is also perceiving (open-ended persons may not plan and structure as much). An introverted user may miss out on communicating details and may answer only the questions asked.

An intuitive user may present general, intended functionalities of the desired system, often in terms of the system's impact on the organization or in some type of abstract synopsis (e.g., "this system must help us to predict the shift in customer base over time").

A sensing user may be more explicit about system details, but may not see enough possibilities for significant change. He may focus on what information he needs now that the current system is not providing, but may need to be guided and given time to consider more impactful system changes or innovations.

A thinking user will provide requirements information in a logical fashion, whereas a feeling user may tend to describe desired features in terms of their easing his jobs (e.g., "the system must provide sales figures broken down by salesperson and compared to the previous month" vs. "the system must provide me with more confidence as to where we are selling well").

A judging user will express requirements with considerable structure and definition, whereas a perceiving user may not be as certain as to what she wants; she may wish to examine a number of possibilities and may be more prone to adding requirements later in the development process.

Also, users may be usefully characterized by the four temperaments, with temperament traits discussed earlier applied to the requirements identification process. However, not only should an analyst be aware of communication biases on the part of the user, but he should also keep in mind his own biases and preferences when interpreting the user's communication. It is not difficult to appreciate the value of a more thorough background in the analyst in type considerations as applied to system requirements determination. Many systems have failed or been only marginally successful because the requirements were improperly communicated or improperly understood. Thus, an ability to extract necessary information from a variety of communication styles, which are largely influenced by personality type/temperament, is not only a desirable, but a necessary competency for the professional systems analyst of the 21st century. However, considerable empirical research must be done and anecdotal information must be gathered so as to build a solid, relevant base for developing such a degree of psychological awareness in the IT field.

Users' personalities also have an influence on how users learn—in this case, how they learn to use a new system. Training methods must be developed for effective learning. Thus the user training process can also benefit considerably from MB type awareness.

As we have seen, the application of MB typing in the IS work area can be varied and truly impactful. Whereas a number of IS departments have provided their staff with an orientation to Myers-Briggs typing, not that many have yet proceeded much farther. At least, more involved applications have not been publicized extensively. Pioneers are definitely needed in this area, people who are willing to research, apply, and publicize more elaborate applications and their impact on the organizations.

Potential Drawbacks

Having examined potential applications and resulting benefits, it is worthwhile to examine possible drawbacks and pitfalls in using MBTI in information systems work. A primary requirement of MB application must be, as in medicine, to "at least do no harm." In what ways might *misapplication* of personality assessment contribute negatively within the IT areas?

Firstly, to be truly effective, typing of individuals must be accurate. What is important is that a person's *true* type be determined, and not, perhaps, a "type" formed by repressing one's true preference to fit the work environment. For example, the IT culture has strongly reflected the thinking process, possibly undervaluing feeling. A true feeling type may have repressed such a function at work because he may have felt that feeling was not sought after, appreciated, or rewarded. Such a person may answer the MBTI questionnaire according to his adapted behavior rather than true preferences.

Secondly, and perhaps more importantly, people may be labeled and thought of only as having work capabilities on the side of each of the four preferences, without consideration as to how strong is each of the four non-preferred components. MBTI labels orientation, ways of perceiving, ways of judging, and preference for structure (i.e., the four dimensions) discretely and not continuously. Thus, on a 10-point scale, a person can exhibit an 8-point preference for extraversion and a 6-point preference for introversion; this person would be labeled an "extravert." Another person can exhibit an 8-point preference for extraversion and a 1-point preference for introversion: this person would also be an "extravert." Moreover, the above example shows a difference in *preference* for extraversion rather than a difference in *capacity*. The same way of classifying, of course, applies to the other three MB dimensions as well. The drawback of the method of classifying used by Myers-Briggs is that the capacity for the non-preferred side of a particular dimension is not specified. To what degree is an intuitive person capable of sensing? To what degree is a perceiving person capable of judging?

If, on a project team, we say we have five intuitives, how *strong* is the preference and capability for intuition in each person? Also, a particular preference should never replace lack of appropriate competence in the core IT skill set and knowledge base. An extraverted, feeling systems analyst will use her personality characteristics truly effectively only if she accompanies extraversion and feeling with systems and business knowledge, as well as tact and proper motivation.

While MB personality traits can be one aspect of a person being considered or a specific role within IT, it should not be the only or the dominating characteristic. One can rely too much on the structure inherent in MBTI and interpret the results in an inflexible manner. This tendency may be particularly present in "STJ" types that form a large fraction of IS professionals.

It is essential to understand that MBTI identifies only some commonalities among persons of the same type. Obviously, persons of the same type will differ considerably on other characteristics. However, MB type consciousness may well be a first step for IS workers to start focusing seriously on psychological factors in their work. After seeing both benefits and drawbacks of MBTI, they may be motivated to develop awareness of other human dimensions such as those outlined in following chapters.

It is also important for the IS professional to develop as much as is practically possible, the non-preferred side of his personality, and to become explicitly aware of how such development can assist his work. One's preference should not become an excuse for a lack of balanced professional and personal development. Another possible area of application where the use of MBTI can have significant drawbacks is in the hiring of new staff. Apart from the ethical issue of asking personal questions (as in the MB indicator), which is discussed later, there is the issue of discriminating against people who have significantly developed the non-preferred side. An introvert according to MBTI may be perfectly able to converse meaningfully with team members, users, and higher management, and to give enlivened presentations to larger audiences. Yet the label of "introvert" might work against such a balanced and capable individual.

Furthermore, if it is known that a particular organization uses Myers-Briggs testing as part of its hiring process, a candidate may anticipate the desired personality traits from a job description and "tailor" his responses to suit the job. In fact, one professor in an American university had students fill out the personality questionnaire in different ways to suit different jobs, to demonstrate how such testing could be manipulated.

In discussing benefits and drawbacks, a question often arises as to the consistency of the MBTI result over time. In theory, one's MB personality will (usually) not change. In persons who take the questionnaire again after several years, about 75% are said to come out the same on at least three out of four dimensions. However, MBTI results do depend on a reasonable degree of self-knowledge. A person who is relatively young (in his 20s), and who has spent his formative years repressing his true nature, may only at that point be "coming into his own." Such a person's MB results may change somewhat as a truer self is allowed to emerge.

Ethical Issues

"Ethics" is defined as "the field of study that is concerned with questions of value, that is, judgments about what human behavior is 'good' or 'bad'" (www.hyperdictionary.com). A prime concern in introducing MBTI in IT work to a degree beyond a one-day informative workshop is that no harm be done as a result. Although some ethical considerations do not appear to have an immediate resolution, it is worthwhile to examine points of concern.

The first question relates to privacy. Since many countries consider it improper, even illegal, to ask an employee about his marital status or opinions on controversial social issues, one may ask how appropriate is it to require a person to fill out the Myers-Briggs questionnaire. The argument goes on to point out that the questions asked therein are not, as such, work related. The counter-argument would stress that certain personality components, such as faculty in speaking (with users or teammates) are indeed a necessary part of the job, just as healthy knees are necessary to play ice hockey. Thus, questions to assess personality are "fair game" in recruitment or task assignment.

In environments where privacy is indeed a contentious issue, it is still possible to use type awareness without the type test. One can, without doubt, ask questions related to each MB component that are *related to IS work*. Instead of an MBTI-proper question such as choosing to go to a party or read a book alone, a question from an IS context can be devised to get at similar information. For example, a person can be asked if she prefers to work alone or on a team, to address the introversion issue. Eventually, perhaps an *actual IT personality diagnostic instrument could be developed* containing only questions with an IT work context, but paralleling closely the questions in the MBTI itself.

Another ethics issue relates to using one's MBTI information in work assignments or promotions ("we really need an ENFJ manager here"). Problems with a strictly direct application have been outlined in the previous (Drawbacks) section, notably that strengths on the non-preferred side are ignored in the MB classification. However, a deeper degree of MB type literacy on the part of IS management can lead people to assess the four dimensions by observing *how people react* in the course of their work. An astute manager will notice the intuition of his designer or the structured orientation of her programmer. Such characteristics will be part of the overall impression that the supervisor has of his subordinate. This impression, of course, will be used in personnel decisions.

It is important that IS managers not "jump on the bandwagon" regarding MB typing, but that they get an adequate appreciation and a balanced perspective. Use of MBTI should not disadvantage anyone. It is essential to remain aware of the fact, at all times, that MBTI characterizes preference or tendencies, and not absolute, programmed, and always predicable behaviors.

Further Research and Development

Most people would agree that application of MBTI has been legitimized in the business management area. Also, having examined highlights of both academic research and practical application related to IT, we can say that MBTI has made some inroads in the IT professions.

With a greater emphasis on "emotional intelligence" in many areas of business, now may well be the time for MBTI to begin making a more significant impact on the work of IS professionals. However, for this to happen, specific efforts will need to be undertaken to enable IS managers to understand the MB-IS link in more depth. This section, thus, provides suggestions and challenges as to studies and projects that would facilitate MB type application to IS at a deeper and more comprehensive level.

There are a number of research questions that could be addressed further by motivated academics:

1. Does the distribution of types in IT ("in IT" would need to be defined) vary across organizations, industries, and countries?
2. Do hypothesized connections between type/temperament and system development tasks prove accurate?
3. Exactly *how* does each temperament tend to carry out main IT tasks? Is the way of working consistent with type theory? How is the "order of preference" evident in carrying out IT tasks?
4. To what degree is MBTI used in IT organizations? What impact has it had?
5. What notable drawbacks (if any) have been observed in application of MBTI?
6. What appears to be the actual impact of MBTI use on productivity, effectiveness, and employee morale?
7. What activities/tasks are being emphasized by development of systems supporting e-commerce that were not evident in basic reporting systems? Is more personality type variety desirable in e-commerce project teams?
8. What are the identifiable stages of application of MBTI in IT in an organization? Does the stage of MB application correlate with degree of "IT maturity"?

In an applied area such as IT, practice often precedes theory. Academics will have more to study as IT practitioners get more involved in using MBTI beyond a basic orientation workshop. Moreover, with the availability of a common communication medium (the Internet) accessible to nearly all IT professionals and academics, there needs to be continuous communication of applications, successes, doubts, and failures. To this end, there may soon be a website dedicated solely to the use of the Myers-Briggs Type Indicator in the Information Technology field. Such a site could allow practitioners to publicize their "happenings" and researchers to get ideas for more involved study. Also, the site could allow researchers to make readily accessible their findings on this topic without the requirement of individual searches.

Apart from academic *research*, there is also a call for specific *development* efforts that would help to motivate more directly the application of MBTI to information technology work. Firstly, there already exists a booklet, published by Consulting Psychologists Press, called *Introduction to Type in Organizations* by Hirsh and Kummerow (1990). For each of the 16 types, it contains information on contributions to the organization, leadership style, preferred learning style, problem-solving approach, preferred work environments, potential pitfalls, and suggestions for development.

Such information is more directly useful to organizational management since it applies type theory to specific, relevant issues. Could not a similar booklet oriented solely to IT work, entitled perhaps *MB Type in the Context of IT Work,* be produced? Based on experienced MIS research and extensive anecdotal evidence (perhaps largely made known through the Web), strengths, weaknesses, preferred ways of addressing major IT tasks, and strategies for being managed can be explicitly outlined for each type. Such a resource would be a necessity for MB personality awareness to develop to a truly impactful level in many IT environments.

Another proposed development could be even more ambitious. This would involve *development of a personality indicator parallel to the MBTI,* but using questions only from an IT work setting. This could take the form of a "sorter," similar to the Kersey Temperament Sorter. A main psychometric challenge would be to show that the estimated "types" from such a sorter correlate adequately with the true MBT indicator. Availability of such an exclusively IT-oriented instrument could eliminate concerns (on the part of individuals or even labor groups) as to the ethics of asking questions irrelevant to the work environment.

For MB type orientation in IT to really become commonplace, more training and clarification material will need to become available. Videos, perhaps downloadable through the Internet, could point out explicitly how types and their order of preference "attack" specific issues across the system development stages. Exercises could be provided and discussed, synthesizing both research results and practical observations. Perhaps now is an appropriate time in the evolution of the IT profession to "rediscover" personality effects and to move MBTI application to a more advanced and more influential level. A dedicated website could be a first step.

FOR THE BEGINNER

While it may be quite meaningful to look at MBTI and appreciate its potential in the IT field, a person who has just been initiated may feel "lost" as to how to proceed after having had his/her type identified. The following may prove helpful.

The final arbiter of your MB type is not the Indicator result—it is *you.* Upon receiving the result of the questionnaire, you can go over each component to see if you agree with it. Remember, what is important is your *true* type, not the way you behave at work, which may have resulted in repressing your natural inclinations. If you can agree on your true type, you have taken the first step. If not, you will probably be able to narrow down your type to three likely choices. You may then wish to examine an MB book, such as the classic by Keirsey and Bates. By reading about each possible type for you, you may be inclined to agree that, yes, "this is me."

Knowing your type, it will be helpful to read one or two books on MB types, one of them relating to work in general, such as the one by Kroeger and Thuesen (1989). You are thus developing deeper awareness of your re-actions and approaches. Having understood yourself better in terms of type, you may wish to examine the types of those around you. You will notice and understand much more about why you "click," why you argue, and why you sometimes just "can't figure each other out."

Developing such literacy must be done slowly, recurringly, and at the "gut-sense" level. As with any new material that is eventually digested, this takes time. Eventually, you will have a "gut sense" of your type, your resulting behaviors, and those of others. You will become interested in "type-watch-ing" (of course, you will want to do this in moderation, and not have it take over your life and all your conversations).

After this initial stage, you should have a decent familiarity with the Myers-Briggs personality system, in theory and in practice. At this point, you are ready to look more deeply into how your type affects your job in the IT area. Firstly, see how you use your MB preference at work on each of the four dimensions. Where does some preference (e.g., intuition over sensing) not have an opportunity to be used? Where are you forced to use your non-preferred side to a significant degree? How does this feel to you? What emotions are generated within you as a result of not being able to work according to your natural inclinations? Be specific. Then, look at your temperament type (SJ, NT, NF, or SP). What are the main needs of your temperament? Are these needs satisfied at work? To what degree?

You may now wish to go deeper and examine the order of preference for your type, discussed earlier. This relates to the order that the function of intuition, sensing, thinking, and feeling "kick-in" for your type. This order is different for each of the 16 types. Examine how this order of preference relates to specific tasks in your IT work. For example, an analyst whose order of preference is intuition, feeling, thinking, sensing, may first focus on the overall contribution that a new system will provide to the business, how it will change the business (intuition). She may then focus on how each department executive is likely to react to the system. She may be looking for areas of disharmony and potential conflict, and imagining strategies to diffuse such conflict (feeling, supported by intuition). After this, she may begin to consider new data and communication technology that would be needed in such a system, but at a high level (thinking). However, this person may never get to—and may actually avoid considering—the details of the new system (sensing function is last in her order of preference).

Having become explicitly aware of your work approach and its relation to your order of preference, you may try to identify difficulties you have had and relate them to your new awareness. Two people with opposing orders of preference—and perhaps having an intense need to "be right"—can generate much disharmony and unproductive energy, simply because of a lack of "emotional work intelligence."

Having introduced this psychological awareness to your mental work framework, you may wish to apply it each time you face a stressful work situation, where your energy is drained. You may come to very revealing insights, and this will lead you to make creative adjustments in your mode of operation. However, it is not advisable to do such development work in isolation.

To this end, work groups can be formed, perhaps on a voluntary basis. A centralized website ("community"), dedicated to Myers-Briggs in IT, can be of immense assistance by providing a forum for exchange of insights and feelings, and by providing an air of legitimacy and motivation for this effort. MIS academics can also monitor this website for issues worth researching. They can ask exploratory questions. IT professionals can feel connected to the academics' research efforts and can look forward to their results.

Having developed and applied a deeper awareness of your type and the types of those around you, and having participated in group exchanges physically or electronically, you will have developed a new sense of empowerment. You may begin to recognize type traits in others simply by observation, especially at work (e.g., "with the way he identified all the detailed data items, I can see his sensing strength"). Eventually, this type of awareness and a new capability to react and relate will become second nature. You will then truly appreciate that "Information Systems is a *socio*-technical field."

CONCLUSION

The Myers-Briggs personality type is one major psychological factor that can come to the aid of IT professionals, particularly in an era of flux. However, it is not the only "tool." One instrument cannot say everything about a person's inner dynamisms. However, because of its acceptability especially in business, the availability of supporting literature and its inherent structure (which appeals particularly to the many structured thinkers in IT), the Myers-Briggs type awareness is a very appropriate starting point for developing psychological awareness within IT. The following chapter adds to personality awareness by presenting yet another classification system with its own insights and contributions.

REFERENCES

Berens, L., & Isachsen, O. (1995). *Working together.* Institute for Management Development.

Ferdinandi, P. (1994). Re-engineering with the right types. *Software Development,* (July).

Hirsh, S.K. (1991). *Using the Myers-Briggs Type Indicator in organizations: A resource book* (2nd edition). Palo Alto, CA: Consulting Psychologists Press.

Hirsh, S.K., & Kummerow, J.M. (1990). *Introduction to type in organizations.* Palo Alto, CA: Consulting Psychologists Press.

Hohmann, L. (1997). *Journey of the software professional.* Upper Saddle River, NJ: Prentice-Hall.

Jung, C.G. (1921, rep.1971). *Psychological types.* Princeton, NJ: Princeton University Press.

Kaiser, K., & Bostrom, R. (1982). Personality characteristics of MIS project teams: An empirical study and action-research design, *MIS Quarterly,* (December).

Keirsey, D., & Bates, M. (1978). *Please understand me.* Prometheus Nemesis.

Kroeger, O., & Thuesen, J. (1993). *Type talk at work.* New York: DTP.

Lyons, M. (1985).The DP psyche. *Datamation,* (August 15).

Myers, I.B. (1995). *Gifts differing.* Palo Alto, CA: Consulting Psychologists Press.

Smith, D.C. (1989). The personality of the systems analyst: An investigation. *ACM Computer Personnel, 12*(2), 12-14.

Teague, J. (1998). Personality type, career preferences, and implications for computer science recruitment and teaching. *Proceedings of the Third Australasian Conference on Computer Science Education* (pp. 155-163). Association for Computing Machinery.

Thomsett, R. (1990). Effective project teams. *American Programmer,* (July/August).

White, K.B. (1984). A preliminary investigation of information systems team structures. *Information & Management, 7*(6).

White, K.B. (1984). MIS project teams: An investigation of cognitive style implications. *MIS Quarterly, 8*(2), 95-103.

Yourdon, E. (1993). *Decline and fall of the American programmer.* Upper Saddle River, NJ: Prentice-Hall.

Chapter II

Enneagram Personalities

INTRODUCTION

Recalling the definition of personality as "a complex set of relatively stable behavioral and emotional characteristics," we can appreciate the insights provided by the Myers-Briggs system. However, we are equally aware that no one system can hope to address all aspects of personality.

A noteworthy personality analysis tool that has achieved a significant presence in both personal growth and management applications is the *Enneagram system of personalities*. The essence is said to have descended from the ancient Sufis, and modern adaptations have been made by a variety of authors, including Riso (1990), Condon (1997), Palmer (1998), Rohr and Ebert (1990), and Goldberg (1996).

Whereas MBTI attempts to explain *how* we function, the Enneagram focuses more on *why* we function in a particular way—what is our underlying emotion that guides the way we act? In this way, MBTI and the Enneagram can be viewed as complementary.

NINE BASIC TYPES

The Enneagram proposes nine personality types, with each type being assigned a specific number, one to nine. It is assumed that each type operates under the often subconscious influence of a specific emotion, as pointed out in Table 12. Moreover, each type is said to have a main desire, along with a

Table 12: The Underlying Emotions of the Enneagram Types

Type	Emotion
One (Perfectionist)	Anger
Two (Helper)	Pride
Three (Status Seeker)	Deceit
Four (Artist)	Envy
Five (Knowledge Seeker)	Greed
Six (Loyalist)	Fear
Seven (Fun Seeker)	Gluttony
Eight (Power Person)	Lust
Nine (Peace Maker)	Sloth

main fear. Such tendencies are often explained by the notion that no one has had a perfect upbringing environment. The interaction of a person's innate emotional "slant" and the type of inadequacy in the upbringing environment are said to give rise to the nine ways of viewing and reacting to reality.

In short, each of the nine types requires some "condition" to be satisfied before he/she can experience him/herself as being "OK." Behavior is guided by the underlying emotion so as to satisfy the critical condition.

The Perfectionist - 1 can feel OK only if he is performing nearly perfectly—if he is doing the "right "thing.

The Helper - 2 can feel OK only if she is helping someone else, and receiving appreciation from the one being helped.

The Status Seeker - 3 can feel OK only if his image in front of others is that of a successful, important person.

The Artist - 4 can feel OK only if she is free to be unique—different from everyone else and to express this uniqueness in her own way.

The Knowledge Seeker - 5 feels OK only if he has impartially observed, analyzed, and learned as much as possible about his environment.

The Loyalist - 6 can feel OK if she is obedient to external rules and dictates; she is often hesitant to act out of personal initiative.

The Fun Seeker - 7 feels OK if he is having fun, even in difficult situations; such an eternal optimist carefully avoids suffering and may overlook important warning signs of impending danger.

The Power Seeker - 8 feels OK if she is the "mover and the shaker," not having to submit to others.

The Peace Maker - 9 feels OK if there is no conflict—often at all cost.

Figure 1: The Enneagram—Showing Directions of Integration

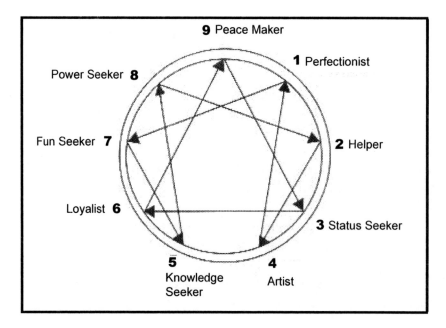

Having looked at the above overview of the nine types, we can undoubt-edly notice people in our personal or work lives who fit some of the above descriptions. While each type is associated with a particular "compulsion" and thus a type of dysfunctional energy, the actual diagram of the Enneagram system (Figure 1) offers what are called "directions of integration" (arrows for each type). Conscious progression for each type along the indicated path allows the type to become broader, deeper, and more balanced, thus utilizing psychological energy more effectively and more dynamically.

Each type also has two "wings," the types of the preceding and follow-ing numbers. For example, Type Two has a 1- and a 3-wing, while Type One has a 9-wing (preceding) and a 2-wing. A person of a particular type will usually tend to favor one of the wings over the other. For example, a Two with a 3-wing will be a helper, but with some tendency towards image seek-ing, whereas another Two with a 1-wing will have some tendency towards perfectionism in the helping.

The nine types are also divided into three "center" categories: head (intellectual) types: 5, 6, and 7; heart (emotional) types: 2, 3, and 4; and gut (instinctual) types: 8, 9, and 1. Thus, the Enneagram identifies categories of thinking, feeling, and acting types. Clearly, all three categories can make considerable contributions to the IT field.

Consider two software engineers, both of INTJ Myers-Briggs type. Although, they may approach their work similarly, with the same order of preference (intuition, thinking, feeling, and then sensing), each may have a different underlying motivation. One may strongly need to appear success-ful and the other may be motivated mainly by the opportunity to help the users. An appreciation of both of these personality aspects can provide an IT organization with more comprehensive insight on the road to "emotional intelligence."

Determining One's Type

The process of determining one's Myers-Briggs type is more quantita-tive and direct. One fills out a questionnaire that is then assessed according to a key. This assessment process has been used and refined extensively, and possesses generally acceptable levels of "validity." The process of determin-ing one's Enneagram type, at this point, appears to be more qualitative and less direct. Although there are so-called Enneagram "tests" found on the Web, such tests appear to have been developed in an ad hoc fashion without coordinated, extensive validation. Thus, such instruments may serve more as indicators rather than identifiers of Enneagram personality. They can help one to narrow down his possibilities. Traditionally, however, individuals have agreed on their Enneagram number after considerable education (reading, workshops, discussion) and introspection.

Some consultants offering an Enneagram workshop have asked people to identify two possible types that they think they may be. Then, with nine separate personality rooms open, people were asked to go into the room of their first choice, to mix with the people there, and decide if they really "fit"; if they were not sure, they would then try the room of their second choice and so on. For some people it may take considerable time to feel confident in identifying themselves with one specific type. However, the Enneagram need not be used with strict, "scientific" rigor to have a definite impact. The key is in identifying one's underlying emotions and possibly resulting dysfunctional behavior patterns that limit one's vitality, adaptability, and resourcefulness.

Relevance to IT

The Enneagram has been promoted quite widely for about 25 years as a personal development tool, and it has merited serious consideration for management application for at least a decade. In 1994 there was an International Enneagram Conference held at Stanford University that attracted 1,400 people. Recent books such as *The Enneagram in Love and Work* by Helen Palmer (1998) and *Getting Your Boss's Number* by Michael J. Goldberg (1996) have related Enneagram theory specifically to work situations. In such literature, passing references to IT work are made. However, there has not yet been a specific, widely publicized effort made to link Enneagram personalities with IS development work.

However, with the broadening scope of Internet-based information systems, IS developers are forming an increasingly non-trivial fraction of the working population across many countries in the world. Thus, it should not seem unusual to take at least preliminary steps to identify potentially significant contributions of Enneagram-based insights to IS work. The initial goal would be to obtain an awareness of underlying emotions and attitudes among different types of IS workers, and to see how such attitudes may be counterproductive.

NINE TYPES DESCRIBED

We now examine each of the nine Enneagram types in some detail and attempt to make preliminary connections to the requirements of functional IT work.

Type One - The Perfectionist

This type of person sees himself as "OK" when his work as well as other major aspects of his life follow high principles and are "perfect" or nearly so. His underlying assumption is, "If you are perfect, then you will survive and be loved." The driving emotion is anger.

It is said that a One, in his upbringing, was accepted and esteemed for his performance rather than for himself. A One, thus has a compulsion to "do what is right" and may often suppress his real feelings by focusing on what *should* be done. Such suppression generates anger, which drives the One further towards his compulsion. This type's false claim is, "I know the right way." He is principled, orderly, but often also perfectionistic ("picky") and self-righteous.

By looking at the Enneagram (diagram), we notice that a One may have a 9-wing or a 2-wing. The One with a 2-wing is more helpful to others in striving to achieve the ideal; he may also be more controlling. The One with a 9-wing may be more relaxed in pursuing perfection; he may also be more detached.

Most Ones, like the other eight types, do not exhibit primarily the noble traits of this type, particularly under stress. Unless a person has done considerable personal growth (e.g., such as that suggested in the diagram itself), he will exhibit both positive and less desirable traits that arise out of limited awareness. While Ones at work can be principled, ethical, fair, and conscientious, they can also often be critical, controlling, inflexible, non-adaptable, and obsessive. They may work hard but may not "work smart."

Perfectionists are likely to be attracted to computing, because of the inherent structure, order, and a "right way," with easily assessable, tangible results. Thus it is likely to find a number of Ones among professional programmers. The area seems to "feed" their compulsion. Moreover, it is quite possible that in programming or other strictly technical areas of IT, the strengths of a One will indeed be desirable, whereas his weaknesses may not become as apparent.

However, as the technical personnel move into positions with a "softer" component, Ones, if unaware, may find themselves considerably handicapped and exerting a less-than-functional influence on their working environment. Goldberg refers to accounting as a "One profession"; this could likely be said of computer programming and of the "IS departments" of the '60s and '70s. Today, however, IS and IT have taken on a broader scope, necessitating the broadening of personal horizons for a One in this field.

As a programmer, the One will undoubtedly focus on quality and efficiency, on following all standards or creating them. Software quality control is another area of excellence for a One. A One project manager is likely to focus on schedules and organizing. He may, however, experience a conflict in a situation where quality of the product must be compromised in order to meet a deadline. He is likely to emphasize security and ensure precautions (such as backups) are always taken to avoid unnecessary risks. The One project manager may give the project a sense of purpose.

The One IT manager will stress high quality, competence, and commitment. Some may be "brutal, but fair." The 2-wing manager may care particularly about developing subordinates' potential in the interest of a high-quality work environment. A One manager is effective where there is a clear reporting structure, unambiguous goals, and well-defined responsibilities.

However, there are a number of situations in the current IT climate that would bring out the weaknesses of a One. A One is significantly uncomfortable in "gray" situations where his work is not easily assessed (since, then, it cannot be easily deemed "perfect"). Strategic IT planning, often open-ended with different possibilities and competing interests, may become overwhelming. A One may overemphasize technical efficiency at the expense of organizational effectiveness of a system. A One may have difficulty in teamwork, where group consensus is necessary. Why should he debate with others if he knows he is right?

A One manager may limit consideration of possible alternatives and thus stifle creativity of subordinates and co-workers of other Enneagram types. Driven by the ideal and exceedingly responsible, he may find it difficult to compromise amidst the ever-present realities in IT environments such as tight budgets, unrealistic deadlines, and conflicting requirements. He will tend to assume that there is one right way to proceed—a One needs to be "right," for emotional security. He may not be as tolerant of learning curves, inevitable human errors, and systems failures.

A One subordinate wants to know exactly what is expected of him; he may become quite distraught with vague directions. He looks to rules, standards, and procedure manuals for guidance. A One analyst may also not allow a user adequate opportunity to express his/her needs in a new system, since he may be impatient to arrive at a tangible requirement list in a short time. He may become too intense and "not see the forest for the trees." A One analyst may tend to overreact to criticisms from his manager or the users, and interpret such criticism as rendering meaningless all of his positive effort.

A One user, however, is likely to be very clear and precise in communicating to the analyst what is required of a system. However, with both the user and the analyst of Type One, possible, feasible, and impactful alternatives may not be considered and assessed in the interest of avoiding ambiguity.

If a One is not aware of his emotional pitfalls, his well-intentioned, idealistic, and sacrificial efforts may yield significantly unfavorable results, which would then generate compulsive criticism, finger-pointing, and a generally dysfunctional environment.

In order to become a more effective IT professional, what is a One to do? To gain practical wisdom from the Enneagram, the One must first accept, intellectually and emotionally, that his image of achieving acceptability and validation through perfection is only an image—he is much more than the quality of his work. Secondly, he should acknowledge that each type operates out of some misconception, compulsion, or at best "half-truth"; thus all types can use further personal growth—all humans, including the One are not (and need not be) perfect. Thirdly, he must be willing to work systematically and patiently at self-development in both professional and personal arenas by following the Enneagram's "growth arrows" and possibly adopting related growth approaches such as those discussed later in this book.

Such a growth commitment on the part of a One IT professional may not be easy to make. If perfectionism, idealism, and excellence in work results have been his ultimate inner security for many decades, what would compel him to change? Change itself, in the IT work environment, plus increasing experience of significant work stress may be the prime catalysts. Of course, an increasing acceptance of personal development/emotional literacy by the IT profession as a whole would certainly be very supportive and encouraging.

There are certain things that the One should and should not do in his initial attempts to break through and break free, rather than break down, due to an underlying limitation in his experience of work and of life. For each type on the Enneagram, there is an arrow leading into the type and an arrow leading out of it. The arrow out of the One points to Seven. This is the *direction of integration* for the One type. The arrow into One comes from Type Four; for a One to move into the psychological space of a Four is a *direction of disintegration* according to the Enneagram theory.

Thus, when a One is "on top" of his work and his other life aspects, he will naturally tend to move to the Seven space, that of the fun seeker and innovator. If, however, the One is under considerable stress (and feels incapable of being perfect), if he is unaware otherwise, he will naturally move towards Four, that is , in the direction of disintegration. For a One this means he will become excessively melancholic and introspective. He will generate feelings that his structural perfectionistic world view will likely have difficulty handling. This may yield to an escalation of stress, which may culminate in a serious emotional breakdown.

If under stress and in considerable discomfort, the One IT worker needs to guard against indulging in too much introspection, self-doubt, and depression. Instead, he needs to make an effort to move towards the Seven space; he needs to break free of the straightjacket of ideals and perfectionism, and allow himself to have fun. In time, the One may even have developed the courage to laugh at his excessive perfectionistic tendencies, perhaps even in front of team members or subordinates. This is the first step in growth for the One, since Seven is the first stop on the path to a One's development as a rooted and well-balanced person.

If we examine the Enneagram itself more carefully, we notice a triangle (with numbers 6, 9, and 3) and a six-pointed figure (with numbers 1, 7, 5, 8, 2, and 4, which points back to 1). This diagram is intended to identify stages in personal (and professional) growth for each of the nine types.

For example, a One at the start may be, at best, average in his perfectionism, leaving him "unfree" and functioning sub-optimally in many areas of life and work. The indicated growth path for the One is: 1-7-5-8-2-4-1, from the diagram. This path would be used for each type except Three, Six, and Nine, but the starting point on the path for each type would be different. For the One, the first move is to a Seven (Fun Seeker); he must "loosen up" before any other growth can occur. After this stage, he should progress towards Five (Knowledge Seeker). The fun and loosening will have awakened the desire to explore, to innovate, to go beyond prescriptions; now the One is ready to learn for the sake of learning, for the sake of deeper understanding. With enough thorough knowledge, the One at Five is ready to move to Eight (Power Seeker). To grow at this point, he should use his capacity for lightness and openness (7), and his gained knowledge (5) to initiate change and to make a difference in the environment, in an independent way. A One at Eight is now challenged to move towards Two (helper). He is free and capable enough to use power and knowledge; now he is invited to add the use of his emotions, for the assistance of others. Once a One at Two type has progressed this far, he is finally ready to move towards Four (Artist), where he can focus on his own inner feelings, unblocking areas that still make him less than optimal. Following this step, the diagram returns to One, indicating that the path through the indicated points has now resulted in an integrated, balanced, and optimally functioning "healthy One." Thus, we have seen how the Enneagram typing system contains an inherent prescription for growth for each of the types.

For the One in IT, the key is to recognize one's limiting behavior as a result of the compulsion to perfectionism, without diminishing one's already positive behaviors associated with this type. Then must come the desire to "outgrow oneself," starting with the move towards fun, imagination, and "loosening the reins." The One must affirm to himself: "I am adequate with all my imperfections." This genuine step will already loosen considerable bound-up inner energy, a scarce resource in today's IS climate, and will likely change the "vibrations" of the One's working environment.

Ones do have a strong commitment to ideals, standards, and quality. It is very desirable to expand the work ideal of a One in IT to include the ideal of an IT worker *who is functioning near his psychological optimum*, while admitting to limitation. If this "vision shift" is achieved (ideally, throughout the profession), Ones may be indeed very ready to glean as much as possible from the Enneagram and other concepts in this book.

Type Two - The Helper

The Helper sees herself as "OK" when she is helping others and expressing positive feelings towards them. She is a heart person and relates to others personally and individually. Her underlying assumption is, "You will survive and be loved when you are helpful"—that is, meeting the needs of others. The underlying emotion is pride. The Two often feels that she is indispensable to the success or happiness of another, that she knows another's needs better than the other person.

It is believed that a Two, when growing up, was accepted and loved for her service to others and not for herself. She may have been an older child given the task of assisting younger siblings. The Two will feel acceptable in her own eyes because of her helpfulness, but may actually lack a deeper sense of self-worth. Her helpfulness arises more out of compulsion than out of conscious choice. A Two's false claim is, "I have no needs; I only care about yours." However, a Two has a significant need to be esteemed for her helping; this is how she is validated as a person. But, this need may be hidden from the person herself. Such a person may even manipulate others into believing they need her help when, in fact, they may not. Then, she will crave genuine "emotional applause" for services rendered. On the Enneagram, a Two may have a 1-wing, where she will help others to be near perfect, or a 3-wing, where she will help them to appear successful.

In work environments, a Two is concerned, warm, caring, sympathetic, and affirming. However, she may also be patronizing, intrusive, possessive, and covertly manipulative, particularly if she has not undergone a process of personal growth. In work situations, a Two will exercise power, but behind the scenes. She may cleverly choose whom to help and when, yet offer the help genuinely and competently. The Two relates to people—in one meeting with a new employee, she may find out more about the person than a One would in an entire year. Furthermore, the new employee will have been glad to have shared so much with such a supportive and open person. The Two knows *people*; in the organization, she is aware of people's feelings and needs, and tries to meet those needs in order to be appreciated. For the Two, success is not primarily a matter of following principles, but of affecting people, be it customers, co-workers, or higher management. In a specific situation, rules may be bent if people's needs and feelings are involved.

Twos are often attracted to the helping professions, although a number of Twos are also found in business. Twos are rarely "bookish" and may seem lightweight where significant linear thinking is required, but they carry considerable weight in accomplishment through people. They focus on people, not on information itself.

Thus, we may ask ourselves if we would find a significant number of Twos in the IT area. Research findings are not yet available on this issue. One may, perhaps, assume that Twos would not be attracted in large numbers to pure computer science. However, there is an increasing number of branches of IT work where helping people (e.g., users) is of significant importance. Study programs in information systems and network support may indeed be attractive to Twos with a degree of technological inclination. Twos with a One (Perfectionist) wing may indeed find the combination of helping and structured results quite attractive. Twos with a Three (Status Seeker) wing may appreciate the status in an organization that a person who is knowledgeable in IT and also people-oriented may achieve. In a preliminary survey of 191 Canadian IS professionals, 12% indicated that the prime motivation for their work is to help people.

In an era where IT is exerting increasingly more influence, the socio-technical nature of this field is increasingly being emphasized. We can truly say that the climate is ripe for Twos in IT. Many promising systems have failed to materialize because of miscommunication with users. IT "techies" have often been accused of not understanding and even not caring for true needs of users. With a few Twos in IT, such situations will improve significantly!

In system initiation/planning, exerting influence may be required to get the support and approval of key management personnel. Supplementing logical feasibility arguments with the personal persuasiveness of a knowledgeable and well-balanced Two type can be invaluable. In fact, such explicit considerations, when based on high ethical principles, may be a sign of an IT organization that is maturing in "emotional intelligence."

During systems analysis/requirements determination, users may: i) be intimidated by the prospect of having to work with a new system, or ii) not know exactly what information they will really need. Recalling that a Two focuses on the person rather than on information, we can see a significant contribution in making users feel at ease and in generating meaningful, open dialogue about actual requirements. A Two can genuinely encourage active, enthusiastic participation of users on joint development teams, which can make the users feel that indeed it is "their system" that is being developed.

In prototyping/visual design, a Two will not be satisfied unless the user is genuinely satisfied. This is particularly significant in the design of websites and influential, high-profile Web-based information systems. A Two's personal touch on a programming team (e.g., in structured walkthroughs) could certainly alleviate anxiety or fatigue. As a project manager, a Two would be inclined to balance schedule consideration with the actual feelings, needs, and difficulties of the developers. A Two's natural domain would be at a help desk, in user training, and quite possibly in network support/administration.

It is also quite likely that Twos will be among the first in the IT field to promote "soft skills." They can be invaluable in helping to develop employee retention strategies by focusing on employees' needs. They could risk implementing alternative work arrangements (e.g., telecommuting) and employee assistance (e.g., nap rooms, meditation rooms, on-site massage therapy sessions, etc.).

To optimize the contribution of a Two in IT, this type of person will need to be well managed from outside and within oneself. The often more technically oriented IT staff, particularly management, will need to realize more explicitly who the department's Twos are, and what unique and valuable contributions they could make, given the appropriate environment. They will need to offer Twos work in areas where their helpfulness and people skills will shine. A manager of a Two must also never forget her craving for genuine appreciation.

The Two must also manage herself and her personal growth. Firstly, if she finds herself in an IT job where helping is minimal, she will need to find the courage to change to a more favorable position. Secondly, it would be wise for her to discover more explicitly to what degree her love of helping is compulsive, that is, done to meet an inner need for affirmation. Thirdly, she may use the Enneagram to note that under stress, a Two moves dysfunctionally to an Eight (Power Seeker). If a Two is not allowed to help or not appreciated for it, pressure will build up and the Two will express control and domination directly, like the Power Seeker type. A Two can become intrusive, wanting to take charge directly instead of through helping. Such situations can arise in IT, for example, at a help desk. A person can be handling too many calls and may not have had sufficient training on the recent technologies to help adequately. Consequently the anticipated appreciation will often be missing.

To combat one's tendency to dominate directly while under stress, the Two can use the Enneagram to identify the *direction of integration* for her type, which is to move towards Four, the Artist. The Four, as we shall soon see, spends an inordinate amount of time focusing on his deep feelings and emotional needs. This is something a typical Two has always avoided, while compensating by helping others. A Two needs to realize and value her own feelings and to learn to satisfy her own needs for appreciation more directly. This process requires humility, to admit to being needy and to accept genuine human support in the neediness rather than to substitute it with receiving appreciation for helping. A more "wholistically managed" IT organization may provide stress support groups, from which a Two could definitely benefit.

Following the Enneagram, a Two's prescription for becoming well rounded and deeply rooted is to progress, in time, through the path, 2-4-1-7-5-8-2. In this way the helper will have "tasted" the orientations of five other types, and will have become more balanced, energized, and functional. The contribution of such a person in IT cannot be overestimated.

Type Three - The Status Seeker

The Status Seeker (Performer) sees himself as "OK" if his image is one of a successful person. His underlying assumption is "being successful and achieving allows you to survive and be loved." The Three's underlying emotion is deceit. Threes have a compulsion to always achieve and to be seen as a success. Their motto is, "If I appear to be successful to important others, then I am successful" (who I am is who I appear to be).

It is said that in a Three's childhood, achievement (for the sake of image) was valued over everything else. Parents may have shown little, if any interest in accompanying them emotionally. Thus, perhaps subconsciously, Threes believe they have little or no intrinsic worth. They will work long and hard in order to achieve a high score in the report card of success. A Three's false claim is, "I do not fail, I always get the job done." He is a pragmatist who invests considerable energy in the workplace so as to accumulate achievements and the accompanying respect. Some textbooks differentiate between two types of Threes, those who genuinely drive themselves to succeed, forcefully but fairly, and those who are so image-conscious that they will allow themselves behavior of questionable ethics so long as their image is highly acceptable.

In the workplace, the Three is goal-oriented, energetic, motivated, competitive, efficient, and may be empowering. He may, however also be calculating, defensive, pretentious, arrogant, and insensitive. For a Three at work, the bottom line is efficiency, effectiveness, and a winning image. Many Threes tend to be "type A" workics ("workaholics") compulsively driven to succeed for the sake of a successful image. A Three often lacks convictions—life lacks meaning outside of a winning image. As Goldberg puts it, "Threes end up *selling* themselves instead of *being* themselves." They have a marketing orientation to work and to life. Often they will motivate others with their high energy and can become role models.

However, Threes can be intense and insensitive. With a Three manager, human factors can get overlooked in favor of success. Threes may abandon their commitments to people once the people are not needed for their goals. Dysfunctional Threes may go as far as unethically forcing co-workers out of their jobs in order to minimize the threat from such people to their image. For them, all this can be rationalized as "competitive strategy." A Three, on the Enneagram, may favor the 2-wing, being more encouraging and popular, or the 4-wing, being more imaginative and pretentious.

How (and how often) does a Three fit in the IT area? At first glance, this type seems to be more suitable to marketing or general management. It is quite possible that not many Threes would like to spend long, solitary hours programming a computer; however, the current IT area is ripe for "success stories." People have become very successful in terms of image and money as a result of imagination and hard work. With the added marketing orientation offered by the Internet and e-commerce, IT is becoming more attractive to image-conscious people (who, 25 years ago would have shunned the mundane world of COBOL-based transaction systems). The CIO is a strategic manager, in the limelight—Threes love this.

The focus of a Three in IT is not so much to have their work be good, but to have it, and themselves, *look* good as a result of it. A Three would be attracted to areas of IT where there is a high profile and possibility of frequent, ongoing recognition of success. He may not be comfortable being camouflaged by a "successful team"; he would need to be the successful team leader, project manager, or head designer. Thus, it is quite possible that Threes in IT may be found more in small firms than in large IT departments.

As a project manager, a Three would do his utmost to avoid a bad image. This may, however, include driving subordinates unreasonably (if "gung-ho" motivation will not work) or cutting valuable system features to bring a project in on time. With a Three, system quality can suffer in favor of a good image. A new product's marketing efforts may overstate the actual product functionality. A higher-level IT manager may commit to unnecessary state-of-the-art technology so as to enhance the department's image. In interacting with users, he will be "interested," to the degree that it helps his image in the organization. A Three's managerial decisions will largely be driven by their effect on his image or his professional evaluation.

Unfortunately, in IT projects, the more ambitious they are, the more likely they are to fail, at least to some degree. IT is often blamed for other organizational inadequacies. Failure is very hard to take for a Three, who has banked his entire worth and life-meaning on a successful image. Thus, a Three will try exceedingly hard to reverse impending disasters, possibly alienating a number of people in the process. Since a Three has mostly disregarded his emotional needs, he may be at a loss as to what to do, how to retain composure in the face of failure. This type may, in such situations, be prone to burnout or substance abuse.

According to the Enneagram, the Three, under stress, naturally moves to a Nine, the oft-complacent Conflict Avoider. He essentially surrenders his drive and experiences not only professional but also personal failure. An IT manager in such a state can indeed perpetuate negativity and lack of purpose among many subordinates.

For personal growth for a Three, the Enneagram uses the 3-6-9 triad. The direction of integration is to move towards Six, the Loyalist. Rather than being motivated by entertaining a successful image, the maturing Three now becomes motivated by genuine loyalty to the organization. Eventually he will end up at Nine (Peace Maker), who is content and balanced no matter what chaos has arisen in the outside environment. At this point, the Three will have developed a personal self, a personal sense of worth, and can then function as an integrated Three working for a genuine image of success and quality, while not needing to rely on such an image for personal validation.

Type Four - The Artist

The Artist (Individualist) sees herself and her life as "OK "if she is continually able to be "unique" and able to continue expressing this uniqueness. The Four dreads being ordinary; many of this type strongly object to the idea of personality typing, insisting they are "unique, individual, and unrepeatable." The Four's underlying assumption is that "to survive and experience love and meaning, they must be unique and special." The Four craves emotional authenticity—to be true to her deep self. Her underlying emotion, however, is envy, as she is often melancholic about not being understood enough or unique enough in her emotional life. She then sees other types having "happier" lives, which she envies. The Four has a compulsion for being "different." I once had a student in business computing who wore one green sneaker and one red one to class; he was also the only male of a project team of eight not to wear a jacket and tie during the final presentation day for a systems development project. This was, undoubtedly, a classic Four.

It is said that many Fours experienced a "devastating dispossession" in childhood after a period of fulfillment. This sense of loss is said to have triggered this focusing on their deep feelings more than any other type. A Four's false claim is, "I am not ordinary." In Myers-Briggs/Keirsey terms, a Four may correspond to the NF temperament. She works mainly from the depth of her feelings, and she is gifted with deep emotional sensitivity and insight. She can gift her environment with creativity, often in a dramatic style.

A Four with a 3-wing is more extraverted and image conscious, while a Four with a 5-wing would be more reserved, intellectual, and perhaps even more melancholy. A Four is inspired, intuitive, sensitive, imaginative, creative, passionate, and personal, but she can also be moody, self-absorbed, intentionally non-conformist, melancholic, self-pitying, despairing, and impractical.

At work, Fours can be passionately inspiring, guided by their deep inner consciousness. They can pick up hidden feelings on a team and are talented in creating products that evoke feeling. They exhibit a characteristic caring about their work environment. In fact, a number of Fours may become leading proponents for this book (hopefully). However, Fours may become easily bored with terse, impersonal logic. They may emphasize feelings too often and too openly, alienating co-workers of other types. They may also struggle to fit in on teams, since they see themselves as unique, not ordinary. At work, a Four employee needs to feel she is special. The emotional needs of a Four must be met for her to function satisfactorily in a work environment.

How do Fours fit in the IT area? Is it worth writing about Fours in IT? No publicized quantitative results exist on the frequency of this type in IT. However, since the Four usually corresponds with NF, it might be safe to say that large IT organizations will inevitably contain a non-insignificant percentage of Fours. One needs to recall that a Four's emphasis on her deep emotional life in no way implies her lack of intellectual/analytic capacity. Such capacity fits in with the entire person, however, in this unique way.

It may also be quite possible that 30 years ago, in the COBOL/transaction system IS environment, few Fours would find a satisfying milieu. Today, however, and even more so tomorrow, the IT field can indeed attract and welcome a significant number of Fours, and it can be truly indebted to them for their unique, impactful contributions.

With the immanent proliferation of multimedia-based systems for electronic commerce, the inspiration and artistic creativity of the Four cannot be underestimated. There is considerable competitive advantage in an e-business system that is unique in appearance and also in functionality. Fours are likely to recognize the needs of users more completely and more authentically. They are sure to ask themselves how the users will *feel* with the new system.

Since Fours are passionate crusaders for what they believe in and voice their beliefs convincingly and authentically, they can be very valuable in business process re-engineering, where they may need to convince entire departments of the benefits of adapting to a new order. In system planning, they can initiate compelling, artistic presentations to the appropriate decision makers about the undeniable necessity of a proposed system. They will, however, need to be deeply convinced of such a necessity. As managers, they are least likely to ignore genuine feelings and concerns of their subordinates. They are also likely to support creativity and self-development among those reporting to them. Indeed, Fours can be catalysts for personal growth among IS professionals, propelling the profession to a new scope and level of professional awareness.

On system development teams, a Four can propose creative ways of solving required problems, be it in the user interface, program design, telecommunications, or other relevant areas. To a Four, building a system will become primarily a feeling experience. If given the recognition by others as a unique individual, a Four may indeed motivate and give rise to more authenticity, trust, and cohesion on a team. Fours can help a team to manage stress by assisting the other types in overcoming emotional denial. Fours can be passionately, dynamically creative in interactive, visual system design, for example, in RAD/JAD settings. In user training, the authentic interest in the user is bound to be felt.

On the "down" side, Fours will truly dislike routine and excessive detail. They will not likely gravitate to system auditing. They may become excessively "wrapped up" in their beliefs about a specific direction for a system's development in spite of compelling logical evidence pointing to an alternative approach.

A Four may not value the real contribution of other types whom they may see as "overly logical" or "boring." If a Four's needs of being seen as a unique individual are not met within the IT organization, she may develop a sense of pessimism and detachment, negatively influencing others. A Four may even become emotionally traumatized in the case of the failure of a large system project in which she fully believed and invested her true self.

Under stress, Enneagram theory indicates that a Four moves towards the Two (the Helper). A Four is normally highly content in her own feelings, but under stress, if her own needs are not satisfied, she will tend to remain in feelings, but project the feelings on others, compulsively helping them to the point of intrusion. The *direction of integration* for a Four is towards One (the Perfectionist), who is ruled by ideals, principles, and not feelings. This perspective balances a Four's "you are what you feel" orientation to work and to life. Fours in IT will not be in the majority. However, their unique orientation can give rise to very valuable contributions in the IT organization of the 21st century. The emotionally intelligent organization will recognize this and will manage Fours accordingly, to everyone's benefit.

Type Five - The Knowledge Seeker (Thinker)

The Knowledge Seeker (Thinker) sees his life as "OK" when he is able to comprehend as much as possible in his environment, so as not to become overwhelmed by it. The Five's underlying assumption is that "being knowledgeable allows you to survive and be loved." The Five is the thinker, the theorist, the model builder, the detached observer. The driving emotion of a Five is greed, not usually for material possessions, but for increasingly more knowledge.

In growing up, the Five may have developed intellectual capacities as a defense against the intrusion of others. Parents may have been overly attentive or possibly non-supporting; the child retreated into a world of knowledge, largely withdrawing from feelings and relationships. A Five lives largely in his mind, and protects his mental space and his privacy with strong boundaries. He is very protective of his time and energy, and wary of the intrusion of others. A Five is very stingy with his time for others.

The compulsion of a Five is to learn, to gather information, and to understand in depth. A Five friend once told me, "One of the few true joys of my life comes from really comprehending something." The false claim of this type is, "I know all about it." Fives excel at developing theories and long-range planning. They prefer to be observers and analysts rather than active players. Many of the seminal advances in pure and applied computing largely originated with Fives.

The Five is thoughtful, thorough, knowledgeable, perceptive, innovative, and self-reliant, but may also be detached, distant, too abstract, and self-absorbed. It is likely that the Fives correspond mainly to Myers-Briggs/Keirsey's NTs, perhaps even more to the introverted kind. Fives at work can be brilliant visionaries, strategic planners, high-level analysts and conceptual modelers, researchers, and architects. These are the people who conceive and propose new paradigms with intellectual depth and rigor.

However, the drawbacks of a Five at work relate largely to his infrequent access to emotions. He does not focus on "people issues" to any significant degree; to him, doing so may disrupt the clarity of his intellectual vision. A Five's intense desire for privacy might make him seem cold, inaccessible, and unresponsive. Fives are known for not answering messages and calls. Also, they often tend to ignore standard operating procedures and administrative "red tape," which they view largely as nuisances and distractions from "the real work." Fives may also show a disdain for people they consider intellectually inferior and may not recognize the wealth of other skills such people offer to the workplace.

A Five with a 4-wing (Artist) is more creative, sensitive, but self-absorbed; a Five with a 6-wing (Loyalist) may be more loyal and hard working, yet anxious and cautious.

It does not take much to see that the IT field is a natural domain for a Five. However, as the field broadens in scope (e.g., as a *socio-technical* field), a Five is likely to excel only in specific IT segments. A Five is a natural researcher and developer of new paradigms. It is likely that structured programming and the object-oriented development were initiated by Fives. Many of the major hardware and software advances came about as a result of committed efforts of Fives, often in seclusion. Also, the IT field is one of accelerated, continuous learning—a natural orientation for the Five.

In systems development, trust a Five to maintain a multi-dimensional strategic vision and to give high-level impetus to systems for competitive advantage. A Five can relate ideas and information in ways previously not considered. Since this type would rather create new systems (organizational and computer-based) than work within existing ones, a Five can make significant contributions to business process re-engineering/improvement, as often required with implementation of Enterprise Resource Planning systems.

In the development of Internet-based e-commerce systems, a Five can contribute with a vision for unique functionality of a proposed system, likely providing for an increased market share or for more effective telecommunication aspects. His creativity may be more conceptual rather than detail-oriented. On a development team, a Five will generate respect by the depth of his insights and the scope of his vision. A Five can be an essential contributor in an IT organization, provided he is allowed to contribute in his preferred domain. Assigning a Five to routine, detailed work or primarily to people management may prove counterproductive to both the employee and the employer.

The Five in IT, a lover of learning, could indeed dedicate himself to learning about his type and about suggested steps for his personal growth. Under stress, Enneagram theory shows that a Five will tend to move toward a Seven (Fun Seeker). If he is frustrated and not able to use his visionary, rigorous thinking talents, the Five may simply entertain himself (e.g., with computer games), further withdrawing from the issues at hand. The prescribed *direction of integration* for a Five is towards Eight (the Power Seeker). This in effect says, "You have done enough ruminating, analyzing, and understanding—now, go and *use* this knowledge actively; get involved more directly in letting your knowledge provide power to your environment". This step, however, requires emotional commitment to "step out" and promote one's expert understanding among influential people. This step will move a Five from simply comprehending to leading and realizing.

As we further refer to the Enneagram for a Five's growth path, we notice a progression from Eight to Two (Helper) and then to Four (Artist). It takes emotional investment to move from Thinker towards Power Seeker. One now has to work *with* people to make things happen. Yet the focus is still on what is to be made to happen (the system, the technology, the new marketing approach). In moving from Eight to Two, the Five is now developing capacity not only to work *with* people, but to work *for* people (as a Helper). This is a major accomplishment for a Five, who initially desires minimal people contact in order to observe and to understand. A Five who has passed through Eight and is now at Two still remains a Five in essence, but the Five orientation is being "purified," i.e., it is acquiring "emotional intelligence."

A key shift for the Five, however, happens when he moves on the Enneagram path from Two (Helper) to Four (Artist). At Two, the Five has become comfortable with feelings for others. At Four, the Five has a chance to enter the world of his own feelings, something which his essential orientation has always avoided. At work, he can, with some enthusiasm, explore his emotional reactions to situations or specific people. He can try to notice underlying assumptions that may be giving rise to such feelings, especially if these feelings are hampering his effectiveness on the job. A Five at Four has matured significantly in the emotional realm. He may find further discovery of the deeper self (as proposed in Chapter IV) not just theoretically interesting but practically invigorating. At this point, the Five will have moved beyond his original motto, "I think, therefore I am."

The Five in IT is an essential contributor in terms of depth, breadth, and rigor of ideas. His contribution is central to continuing innovation. Neither the employing organization nor the individual himself, however, need to be complacently content with a limited, though significant role for a Five. With ongoing personal development, quite possibly initiated by the Enneagram insights, the Five in IT could become more balanced, well-rounded, and even more influential. Organizational support of such personal growth could indeed contribute to the retention and flourishing of an indispensable human resource.

Type Six - The Loyalist (Trooper)

The Loyalist (Trooper) sees herself as "OK" when obeying the rules of the environment and serving loyally in an environment created by others. Her perceived self-worth and her feeling of safety come largely from adhering to clearly defined rules. The underlying assumption of a Six is, "Being obedient and loyal allows you to survive and be loved." The often hidden, but driving emotion of the Six is fear. Consequently, worry and anxiety are constant companions of a Six. The Six craves safety and emotional security.

It is believed that, as a child, this type could not show openly what she really felt or thought. She did not experience adequate, open relationships with caregivers. Upon opening up, she may have been judged or ignored (an indirect judgment). Thus, the ability to experience oneself as intrinsically valuable was hampered. In parallel, obedience was likely demanded and loyalty exalted. The inner security of such a type, therefore, depends largely on satisfying the requirements of one's environment dutifully and loyally. Such a type will "dot all i's and cross all t's"; she is probably the least likely to ignore brackets and commas in a computer program! For Sixes, opening up with feelings or unique, deep ideas may be especially frightening—they cannot trust easily that others would accept them. However, being seen by others (and by oneself) as a "loyal trooper" following clearly defined, unambiguous rules is much safer. Even then, many Sixes may still experience a nagging anxiety regarding being acceptable, especially by authority figures, whom they may have difficulty trusting.

In terms of focusing very intensely in order to do a good job, Sixes are similar to Ones (Perfectionists). However, Ones "know they are right," while Sixes "want to be right" but often question whether they are "right enough" in the eyes of those who count, fearing abandonment or humiliation. While it is desirable for all types to obey sensible laws and directions, this obedience is much more central to the emotional existence of a Six. A Six does not tolerate ambiguity well and may be very cautious in embracing change. She may act as a "devil's advocate," looking for a downside in areas of endeavor to which she intends to be loyal. (If being loyal to or "hanging one's emotional existence on" a cause or project is central to one's security, one better check out potential pitfalls as thoroughly as possible.)

The claim of a Six is, "You can always count on me," but the inner root of such a claim is emotional dependence on being dependable rather than a free, conscious choice to commit one's energies. The Six compulsion is to establish emotional security by obeying rules and dedicating oneself wholeheartedly to structured, clearly defined undertakings.

The Six can be a very good team player, showing exceptional dedication and loyalty. She is comfortable and mobilizes energies quite readily in an "us" against "them" situation (users vs. analysts, IS vs. other high-level management). The experience of, "We're in this together, come hell or high water," is particularly energizing for a Six, but even this may not be without doubts about the trustworthiness of others on the same side.

Sixes are loyal, responsible, committed, trustworthy, faithful, and obedient, but may also be excessively cautious, anxious, dependent, indecisive, and contradictory. Sixes with a 5-wing (Knowledge Seeker) tend to be more introverted, intellectual, and cautious. They seek knowledge in order to "be prepared." Sixes with a 7-wing (Fun Seeker) will attach themselves to a cause, working at it in a more extraverted, active, and possibly impulsive way. Sometimes, such a Six can camouflage inner anxiety in the form of somewhat excessive enthusiasm.

Because of their loyalty to the outside without a firm, personal rootedness on the inside, Sixes can be taken advantage of. However, when they can no longer tolerate abuse of their dedication, they may react aggressively. They have difficulty assuming a proper stance in the "box of self-assertedness," and will either "kneel obediently behind the box" or "stand defiantly in front of the box" if they feel their loyalty abused.

At work, Sixes are dedicated, loyal beyond the call of duty. Having a sense that "life is hard," they can come alive under adverse conditions. They work best in jobs where the rules and responsibilities are clearly defined, and where the results of their work can be objectively evaluated. However, Sixes can be so taken up with analyzing components of a problem that they lose sight of the problem itself—they want to have all angles covered "just in case...." Being cautious, Sixes are likely to say "no" to a new idea or proposal. At the very least, they will carry out an extensive analysis to identify potential risks of a new endeavor. They can, thus, be valuable troubleshooters.

Sixes at work are emotionally cautious and reserved, especially with respect to trust of supervisors and co-workers. With them, trust needs to be established consistently over a period of time. Because of their external locus of control, they may overvalue the power of authority. Because of their need for reassurance, Sixes may become excessively emotionally dependent on their work and may have difficulty relaxing, especially if clear, unambiguous feedback as to the correctness of their work is not readily available.

The IS field should be especially attractive to the Sixes. Programming and other technical areas contain no shortage of prescribed procedures. The feedback on the work can be clear and can be experienced as a reward for one's loyalty. Since the Six type has been linked to Myers-Briggs STJ and SFJ, it is quite safe to assume that indeed many Sixes are found among IS workers.

In an IT environment, Sixes want "to do it right." They rely heavily on accepted "tried and true" procedures and approaches. They work step by step, meticulously, and with dedication. It seems that a natural area for Sixes in IT is programming. Loyalty to the rules of the language is rewarded with a working program. Maintenance programming offers a Six the chance to correct errors and to feel a connection to the welfare of the organization by adjusting the system to make it more effective. Sixes can also feel at home in IS auditing and quality assurance.

In systems analysis, Sixes may be somewhat baffled by conflicting requirements, changing scope, and political maneuvering. They may overlook new possibilities because of anxiety over the lack of clear structure and direction. However, Sixes can make a major contribution in analysis and general design through their "devil's advocate" role. They can see the downside and questionable future implication of proposed system alternatives. Such types can also provide valuable insight, from the point of view of caution, in business process re-engineering, as may be needed in ERP implementations.

In detailed system design, Sixes are likely to be thorough, following accepted, prescribed approaches. Their loyalty will generate a responsiveness to user needs. Their caution, however, may miss out on opportunities for competitive advantage.

The "trooper mentality" of Sixes can make them loyal team members once they can develop trust. They can draw a team together in times of adversity with their loyalty and determination. However, they may feel quite threatened by abrupt changes in project direction or by departures from standard operating procedures. Their fear may then drive them to accuse others and to doubt the others' genuine motives.

As IT managers, Sixes provide loyalty, but also demand loyalty. They may be threatened by excessive originality and creativity (as exhibited by Fours) or promotion of untried paradigms (as may come from Fives). When approaching a Six manager with a new idea, it is wise to have studied possible downsides and to give them adequate weight in the presentation.

Sixes are the traditional "troopers" of the IT field. In the older days of predictability, stability, and somewhat limited IS influence, Sixes could have easily found a home in IS for their mode of operation. In this 21st century, IT keywords are change, business impact, and creativity. Such a climate could give rise to significant stress among Sixes. What they are loyal to is changing rapidly. Breaking new frontiers is rewarded more than loyally following accepted procedures. The scope of IS work is expanding to include "gray" areas. The Six is one type in IS that certainly needs to expand his horizons. *It would indeed be wise for the Six to expand his/her emotional literacy.*

In Enneagram theory, a Six under stress becomes driven by fear, the underlying emotion, towards Three. She commands herself, "I must have this done by the deadline." When this appears impossible, she panics and at least wants to make it *look like* the work has been done (the perspective of the Three-Image Person). She may generate voluminous graphs and handouts to camouflage an inadequately prepared feasibility study. She may document an aging system excessively, after not having had the time to find maintenance errors, arguing that she must understand the system first to find errors in it. Such "cover up" work is often done under emotional stress and, if prolonged, can generate questionable work output, mismanaged time, even poorer productivity, and eventually an unhealthy employee.

The *direction of integration* for a Six is towards Nine (the Peace Maker). A Six at Nine begins to adopt a more laid-back attitude about what simply could not work out. She can accept, emotionally, that she simply couldn't have time to prepare the kind of feasibility report that she would have liked to. She can say to herself, "That's the truth," and find deep comfort and inner strength in this. She will not need to generate frantically volumes of "cover-up" material (as does the Six at Three). Such a shift to a Nine is not natural for a Six, however, and will require considerable effort in personal development.

While the loyalty of a Six is a definite strong point, there is evidence that IT professionals in general, and likely Sixes among them in particular, are more loyal to their technology than to their employer. Corporations concerned about IT employee retention can become aware of the Six's need to be trusted and the Six's particular ability to anticipate undesirable situations. In turn, MIS researchers can investigate how Sixes in IT cope with required, inevitable change as compared to other Enneagram types.

A key issue for Sixes in IT is coping with the "new realities" of constant change and increasing scope. The IT Six will need to develop more strength within, as a person and as a professional. She will need to learn that it is OK to take risks and to make mistakes. He will need to learn to trust change and to work with very different types of people. Sixes cannot afford to let their fear and anxiety hold back their significant talents. For the numerous Sixes in IT work, emotional literacy is particularly important. In fact, it is a critical success factor.

Type Seven - The Fun Seeker (Enthusiast)

The Fun Seeker (Enthusiast) sees himself as "OK" when he is not dealing with seriously painful situations and he is "having fun," enthusiastically looking at and promoting the bright side of life. His underlying assumption is that "being happy and connecting with many external stimuli that give an experience of 'fun' makes life worth living." The driving emotion of a Seven is gluttony, not specifically for food, but for more and more "fun" experiences.

It is believed that in childhood, a Seven was, at first, quite happy. Then an event or series of events may have occurred where the world was becoming too painful to handle. The Seven simply denied this emerging reality, went back to "play and fun," and has remained in this orientation ever since. Thus, a Seven lacks interiority—the deeper, "soulful" self (which is the natural domain of the Four).

The Seven type is the eternal optimist, getting involved in increasingly new endeavors in order to "have fun." His compulsion is to avoid pain and boredom in life, which often results in avoiding serious responsibility and pressure. The false claim of a Seven is, "I am always happy." An average functioning Seven is enthusiastic, exuberant, spontaneous, playful, and lively, but can also be hyperactive, superficial, escapist, and unfocused.

The Sevens at work love conceiving and marketing new products, especially those that have an element of "fun." A Seven with a 6-wing (Loyalist) will be more responsible and loyal, but also anxious about having to endure times when fun is largely absent. A Seven with an 8-wing (Power Seeker) may be more exuberant, extraverted, aggressive, and competitive. Sevens generate many ideas for new possibilities, seeing new angles readily. They express their insights with enthusiasm and get people to cooperate by their exuberance and playfulness. They work intensely in spurts, especially when they feel they are "having fun."

Sevens detest closure, tight deadlines, and being "pinned down." They may overfocus on their own capabilities and may thus take on too much. However, they may lack the staying power to follow through on their insights and may become very stressed when their "fun" work is no longer fun. The Seven may have difficulty with commitment and loyalty. When he is no longer having fun, it's time to move on to a new, fun, and exciting situation.

Is there a significant number of Sevens in a serious, exacting area such as IT? According to my preliminary survey of Canadian IS professionals, 19% of the respondents reported that they are motivated to work because "work is fun." Computing, with its mental, puzzle-solving characteristics and rapid, accurate feedback, has—like mathematics, chess, and other cerebral endeavors—been a source of a particular kind of "fun" for some time. Today, with the addition of visual development tools, varied graphics, multimedia, and impactful e-commerce systems, the IT field can increasingly provide a source of fun for the Seven life-view. There is fun in the work itself and fun in seeing the impact of the work on the changing business world.

However, business computing still does require considerable "grunt work" and staying power, especially at the lower levels. It may be difficult for a Seven to follow this path in order to get involved in more "fun" IT activities. However, tools such as visual development environments and CASE can eliminate significant monotony even at lower IT levels. Also, organizations that promote people from user departments to systems analysis positions may provide a more "fun" way for a Seven to join the IT organization.

The main contribution of a Seven in a systems development environment is his enthusiasm and optimism. In situations where deadlines often seem unrealistic, users often change their minds about requirements, and technology does not work as expected, a person's sense of "fun" can be contagious and sustaining. Thus, a Seven, with, of course, a professional sense of competence and responsibility, can be a very motivating project manager and supervisor. However, such a type may be experienced as too emotionally shallow by sensitive subordinates, particularly Four types. A Seven analyst could achieve superior communication with users during requirements analysis. During user training, he would, perhaps subconsciously, be convincing the user that working with the new system will really be "fun."

It is quite likely (although this has not been researched) that Sevens can make significant contributions through vision and creativity in e-commerce applications. Increasing competitive advantage through a Web-based, "colorful," innovative application can indeed be experienced as fun. Sevens could sell such a concept enthusiastically to the organization's management and, if given approval, can motivate systems developers with their optimistic enthusiasm. It is also quite possible that Sevens with a 6-wing will get more fun out of the development work itself, while Sevens with an 8-wing will more enjoy conceiving and promoting the system concept. Of all types, Sevens may be the most willing to embrace change, a constant reality in IT.

The downside of a Seven's initiating and managing a major, impactful project may be his lack of staying power. He loves the conception and the selling of the system, but he is willing to go only so far in enduring hardships to make the idea a working reality. According to Enneagram theory, a Seven under stress moves, mostly subconsciously, to a One (Perfectionist). Unable to have fun in the work yet needing immediate gratification, he develops tunnel vision in an attempt to get back to an enjoyable state. He then controls, judges, and demands, insisting on the one right way to do things. The *direction of integration* for a Seven is to move to Five (Knowledge Seeker). A Seven at Five learns patience and discipline by going deeper into solid knowledge. He can observe and analyze both the positives and the drawbacks in a proposal, in a tight situation, or even in his own work attitude. He can develop capacity for a long, focused effort. Keeping this in mind, one wonders if the exacting "grunt work" in applied computing may, by its nature, draw the novice Seven towards his intended direction of growth. If this should be so, such Sevens, having spent time at and assimilated key orientations of the Five, can then be more integrated once they rise to more influential positions in the hierarchy of their IT organization.

Since a downside of a Seven is his lack of strong capacity for loyalty and deep commitment, this may be an issue for those concerned with retention of IT staff. How can an organization provide a valuable Seven in IT with a win-win environment where he generates enthusiasm to motivate others in his area? With the increasing scope and impact of business information systems, there will undoubtedly be more opportunities for "grounded" Sevens. An emotionally intelligent IT organization will realize this explicitly.

Type Eight - The Power Seeker (Boss)

The Power Seeker (Boss) sees herself as "OK" when she is strong and in control. She believes that "having power and never showing vulnerability allows one to survive in the world." The driving emotion of an Eight is lust, in the sense of taking what you want, of immediate gratification on a grand scale, of craving for satisfaction. The Eight needs experiences of higher intensity than other types in order to feel alive. Eights are not driven as much by achievements as by "potency"; they enjoy the exercise of power. They like to be respected for their significant position of power, and are not afraid to take on considerable responsibilities. Eights emanate a "raw" direct energy and can be quite impulsive.

It is believed that, while growing up, the Eight had to struggle against injustice and possibly abuse; to survive, she had to fight back, repressing her sensitivities. Thus, she got to see the world as a game of power. Often she may use the power to right legitimate injustices. The false claim of an Eight is, "I am always powerful and in charge," and the compulsion is to be independent and in control.

An Eight is powerful, strong, determined, confident, and forceful, but may also be intimidating, excessively controlling, confrontational, explosive, and vengeful. At work the Eight type excels at entrepreneurship and "empire building." She finds it difficult to work in moderation, working mostly in "high gear." The Eight prefers to make gut-level decisions, not taking much time for formal analysis; she usually manages by decree.

An Eight with a 7-wing (Fun Seeker) will be more extraverted and enterprising, while an Eight with a 9-wing (Peace Maker) will be more mild-mannered and quietly strong.

It is not known if there are many Eight types in the IT area. In the IS shop of 25 years ago, a job such as a COBOL programmer or systems analyst would hardly be a primary choice for a lustful power seeker. In my recent, exploratory survey of Canadian IS professionals, only 3.4% indicated they are motivated in their jobs because they get a feeling of power. While there can be a sense of power and control in dominance over the machine, Eights would likely require to feel power within the organization itself. They are not prone to "hiding in the corner" and making a machine jump.

However, the role of IT within many business organizations is increasing significantly. Being able to implement enabling technology for critical business activities does create power. The power is aptly supplemented by the broad organizational orientation an analyst develops while working on different system projects. Thus, it is not unreasonable to believe that more Eight types will indeed be attracted to IT in the near future.

The main contribution of the Eight in IT would be in the management area. If an organization wishes to increase the profile of IT, it would be wise to hire an Eight CIO. An Eight could be a very effective champion for pioneering new technologies or paradigms for system development such as object-oriented analysis and design. An Eight would likely be very effective in ensuring adequate fund allocation to the organization's IT budget. In managing projects, an Eight could be quite demanding on both herself and her subordinates. However, her sense of enjoyment of the power in her work could be contagious and a positive motivating factor. It is surmised that systems development productivity could increase as a result of management by a well-adjusted Eight.

On a development team, the Eight would want to lead, and in a way that generates respect; in IS this would involve both competence and confidence. In times of staff downsizing, subordinates could count on the Eight to do what is possible to save their jobs, but if staff cuts became inevitable, the Eight would not likely agonize in letting people go.

The downside of an Eight in IT is her raw energy and overdrive. Firstly, this work approach could alienate some of the more sensitive, reserved types that are often found in analysis or programming. Secondly, a prolonged "all out attack" approach could risk burnout in an environment of often excessive demand. Under stress, the Eight goes to Five (Knowledge Seeker) according to theory. Instead of going "full-tilt," she will sit back and analyze what is going wrong, shaken by apparent loss of power, which is unfamiliar. Such unproductive ruminating does not regenerate power, resulting in serious stress and lost effectiveness.

The *direction of integration* for an Eight is to move to Two (Helper). Instead of enjoying power for the sense of control it gives, the developing Eight learns to use power to help/empower others. In IS this may mean focusing more on helping subordinates, end-users, or the organization as a whole. To move towards Two, an Eight needs to involve feelings, which may have long been repressed; she needs to regain the child in her that she never was. In further growth, the Eight will move through Two to Four (Artist), where she finds the desire, time, and courage to deal with her own true emotional needs. An increasingly powerful, influential area such as IT can indeed benefit from powerful leadership. The well-integrated Eight can naturally fill this role.

Type Nine - The Peace Maker (Conflict Avoider, Mediator)

The Peace Maker (Mediator) feels "OK" when he is peaceful and not involved in conflict. His underlying assumption is, "Always being peaceful and not rocking the boat allows you to survive and be loved." He displays a laid-back attitude, not investing intense emotional energy for a particular outcome. For him, it is easier to adapt to a not-so-desired outcome rather than to expend considerable energy trying to ensure the desirable outcome. The Nine suppresses his feeling of desire, often using phrases such as "it's only life" or "it doesn't much matter." The driving emotion of a Nine is sloth.

It is believed that, in childhood, a Nine needed to compromise his own desires excessively in order to stay connected to important others. He may have been caught between warring parties or felt overlooked. To avoid conflict from wanting his own way, a Nine learned to give in to others who may have had differing views not only from the Nine, but from each other. As a result, a Nine has superior abilities in mediation between conflicting parties, since he is interested primarily in achieving conflict-free consensus rather than in a particular outcome.

However, a Nine, by placing avoidance of conflict as the main priority, has often given up his own views, strong feelings, and plans; in essence, he has given up himself to make peace with others. Yet, in this, a certain essential life energy has been blocked. A Nine's false claim is, "I am always easygoing and content." His compulsion is to be calm and relaxed.

An average functioning Nine is unpretentious, easygoing, supportive, patient, accommodating, peaceful, and reassuring, but may also be passive, neglectful, complacent, resigned, and lazy. The Nine with an 8-wing (Power Seeker) will be more outgoing and assertive, and the Nine with a 1-wing (Perfectionist) will be more orderly, critical, and compliant.

At work, the Nine thrives on collaboration and excels at generating harmony on a team. He does not like to micro-manage. He proceeds slowly and is not easily influenced by pressure. The Nine does not like to be hurried. He can see the truth, with integrity, in both opposing positions and has the natural tendency to mediate. The Nine takes his time in making a decision while ruminating and viewing the situation from multiple angles. This is not always advantageous; however, because of their apparent ambivalence as to specific results, Nines are not as prone to work stress. Nines may, however, over-emphasize team harmony while stifling or discouraging initiative and imagination.

Much has been written about cohesive, functional systems development teams. Yet conflict between team members, on some level, is almost inevitable. A Nine can make a major contribution here by validating the views of conflicting parties, leading to harmonious compromise.

If, indeed, many STJ types in Myers-Briggs are also Sixes on the Enneagram, this would indicate that the IS field has attracted many Six types. Yet, the direction of integration for a Six is to move towards Nine. Thus the Nine, by being relatively well-functioning, can help to alleviate the fear of many Sixes, especially in stressful situations, which are not infrequent in IS work.

Because of their innate talent for mediation, Nines can be effective JAD leaders, promoting consensus among all parties involved in the joint and rapid development of a business application system. They can also tone down conflict between developers and users regarding feasible system requirements or between those proposing alternate ways to re-engineer organizational processes before ERP system adoption.

Nines can contribute harmony and empowerment to the members of a development team. They will have the ability to affirm dedicated subordinates in trying times. Such affirmation and support may indeed diffuse pent-up energies and allow the other personality types to move in their required direction of integration. A Nine can give subordinates a sense of being genuinely heard and deeply understood.

With their talent for mediation and their laid-back approach to life ("it's only life"), do many Nine types become attracted to the work of IT? There appears to be no publicized answer to this question. However, Nines have risen to prominence in various walks of life. Both Dwight Eisenhower and Pope John XXIII are identified as Nines in Enneagram literature. There is no reason to assume that IT may be too challenging intellectually for an easygoing Nine. However, it is known that Nines do not like pressure: they do not enjoy being pushed. With pressing deadlines, conflicting demands, and ongoing need to adapt to change, Nines may, after some time, "feel the heat." According to Enneagram theory, the Nine naturally moves towards Six when under stress. He begins to worry; he wants to examine and reconcile opposing sides of a problem, but there is no time. He fears catastrophe and doubts his abilities, moving in the direction of breakdown.

The recommended *direction of integration* for a Nine is to move to Three (Status Seeker, Achiever). The complacency, inactivity of the Nine, rooted in sloth, is replaced by action, with the intent to achieve a result. The Nine is then efficient and effective.

It is a challenge to IS professionals worldwide to identify the Nines among them. Notice their ability to dispel unhealthy intensity and their talent for helping to resolve conflict. Such a talent is unique and valuable in the quest for effective and harmonious harnessing of creative energy in an influential field.

Summary

With this brief introduction to Enneagram personalities, it is useful to examine a tabular summary of all the types and their strengths and weaknesses, albeit mostly hypothesized, as related to IS work (see Table 13).

Practical Relevance

Having been presented with an introduction to Enneagram types and their likely behaviors within IS, one may, at this point, engage in healthy skepticism as to the practical relevance of this insight.

Most of us likely know at least a few people who are "classic fits" with some of the Enneagram types. To emphasize this, the following example, taken from the book, *Beginning Your Enneagram Journey* (Brady, 1994), may prove useful.

Ten people are going to an event together. They are walking toward the entrance of the place where the event is to be held. Suddenly, one of them trips on a cracked sidewalk and falls. Each of the others (assuming each is of a different type) will react differently to this experience.

One: "It's inexcusable to leave a sidewalk in such poor condition."
Two: "Awwh. You poor thing, let me help you up."
Three: "Here. I'm good at this sort of thing. I'll do it."
Four: "I fell like this once, and I was in bed for two weeks afterwards."
Five: "Isn't it interesting how everyone is reacting differently to the same event?"
Six: "Are you OK? Oh gosh, now we are all going to be late. This is awful. What should we do?"
Seven: "Wow! What a fall! Hey, but you'll be up and dancing in no time."
Eight: "I'm going to get hold of the building manager and demand that this be rectified immediately."
Nine: "Calm down, everyone. Everything is going to be alright. Let's not get too excited." (Source: Beginning Your Enneagram Journey, *Brady, 1994)*

Table 13: Enneagram Types in IT—A Summary

	Compulsion	Emotion	IT Strengths	IT Weaknesses	Growth Direction
One (Perfectionist)	to follow the highest principles, to be "perfect"	Anger	programming, quality assurance, testing, auditing, technical support, telecommunication	strategic planning, evaluation of system alternatives, group consensus, dealing with ambiguity	to Seven - relax the need to perform perfectly, have fun
Two (Helper)	to help others, with expectation	Pride	system requirements determination, prototyping, help desk, user training	isolated work, data modeling, systems programming	to Four- focus on your own emotional needs freeing yourself from need for admiration by those you are helping
Three (Status Seeker)	to promote and maintain one's image of success	Deceit	team leader, IT manager, CIO, high-profile technical specialist	detailed "grunt work" with a low profile	to Six - focus on working out of genuine loyalty rather than to maintain an image
Four (Artist)	to be always unique and different; to avoid the mundane	Envy	multimedia, website design, visual system development, user manual production	auditing, maintenance programming, prolonged, procedural programming	to One - look outside, to high standards rather than mostly within yourself
Five (Knowledge Seeker)	to seek increasingly more knowledge, mostly for its own sake	Greed	strategic planning, high-level process and data modeling, high-level telecommunications design, research and development, business process re-engineering	people management, user interaction, routine and detailed work	to Eight - step outside your own mind to commit knowledge to powerful action
Six (Loyalist)	to seek security in loyalty and obedience to external parties	Fear	programming, auditing, data modeling, lower level design, testing	dealing with ambiguous situations, accepting new paradigms, conflicting goals	to Nine - don't let fear drive you, don't make goals so absolute- everything is not "a big deal"; seek peace
Seven (Fun Seeker)	to have fun, in whatever situation	Gluttony	system initiation, particularly e-commerce systems for competitive advantage, management, embracing change, user interaction, visual design	procedural programming, lower-level modeling, auditing, work with prolonged isolation	to Five - seriously learn a few things in-depth, even though it may not always be fun
Eight (Power Person)	to be invulnerable, powerful, in control	Lust	management, CIO, champion of change, ERP consulting	procedural programming, low-level modeling, auditing	to Two - use power not to be in control but to help others genuinely
Nine (Peace Maker)	to avoid conflict, to seek harmony at all cost	Sloth	team leader, JAD leader, programming, system maintenance	technical support, work with tight deadlines	to Three - instead of compromising results for the sake of peace, work for successful results

Each type has provided a different response as a result of having viewed and experienced the event from his basic life-view.

From the Enneagram, we can appreciate much more clearly and directly how different people can be doing the same IT work, but from different underlying motivations. Thus, they will recognize different challenges in the same IT work and will exhibit different reactions. For example, in a situation of considering significant reduction of intended project scope, a One may be motivated by meeting the intended original deadlines. A Two, on the other hand, would consider above all the effect of the intended reduction on users. In deliberating whether to introduce a new system development methodology or technology, or possibly whether to change to an ERP system, an enthusiastic (possibly overly optimistic) Seven might favor "jumping right in," while the fearful, cautious Six would likely advocate further, detailed, feasibility analysis, possibly hoping for a "no" decision. In a power struggle between IT and an influential user department related to the direction and pace of automation, the Eight IT manager might adopt a "bulldozing" approach and force the user to accept his decision, whereas a Nine could be overly accommodating to the detriment of optimum effectiveness.

Within the IT employee himself, the key question is how might his reaction drain his own inner mental and emotional energies, making such energies unavailable for solving the problem at hand. Between individuals, how might their differing reactions to a situation they face in common initiate caution, mistrust, and lack of synergy, instead of the intended effective, synergetic communication? Whether alone or with others, a person who has not integrated the various perspectives of the Enneagram to at least a fair degree will experience a considerable amount of stifling energy blockage, under-utilizing his potential and reducing his possible level of fulfillment. Can the IT area of today and tomorrow afford to accommodate passively such a reality?

DRAWBACKS OF THE ENNEAGRAM

The Enneagram is, above all, a behavioral *model*. As such, it cannot explain all human behavior. A common question relates to whether there indeed is only one type for each person, or can a person be a "combination" of two or three? Although theory does point to one predominant type per person (while considering wings), in practice, people, including Enneagram instructors, have at times expressed some doubt. Types 5, 6, and 7 are the head (thinking) types; 2, 3, and 4 are the heart (feeling) types; and 8, 9, and 1, the "gut" (acting) types. Some have proposed that while a person may identify strongly with a particular, say, head type, he may also identify with a specific gut type, and possibly even with a specific heart type. As well, there is not one exclusive and determining psychological instrument to determine one's Enneagram type. The Web offers some "tests," but these appear to provide the role of indicators rather than determinants. While the Myers-Briggs Indicator has satisfied a significant number of psychometricians with its validity and reliability measures, such measures have not been as convincing regarding the Enneagram. Author Jerome Wagner did, however, produce a PhD dissertation on "Reliability and Validity of the Enneagram Personality Typology" (1981); he also published on the topic in the *Journal of Clinical Psychology* (Wagner, 1983). Many of the individuals offering Enneagram workshops, however, do advise participants to acquaint themselves with the nine types from several books, to observe themselves, and to gravitate towards a likely type, which can then be confirmed or changed over time.

In addition, some people do not appreciate the seemingly "negative focus" of this system. The identification of compulsions for all types, some would argue, does not build up one's self-image, by focusing on his shortcomings. Also, while the Enneagram does prescribe a "direction of integration" for each type, it may be difficult for individuals to move there, especially when they are under stress and need it most. Is willpower sufficient, or is the situation similar to that of a person trying to break an addiction, where one is essentially powerless and must appeal to a "higher power"?

The Enneagram may appeal more to those who like structure, of which there are many in IT, since the diagram is in itself a structure for personal development. In any case, the Enneagram does focus attention on hidden emotions arising out of early unfulfilled needs. These emotions are said to drive one's motives, choices, decisions, and reactions. An understanding of one's limiting behaviors and pitfalls, perhaps taken from several theoretical types along with prescriptions for avoiding them, may in itself be a significant contribution of the Enneagram, despite some acknowledgeable drawbacks.

ENNEAGRAM AND MYERS-BRIGGS PERSONALITIES

Enneagram and MBTI typing systems can be viewed as related and complementary. Myers-Briggs identifies *how* one functions in and relates to his environment—in what order the functions of sensing, intuition, thinking, and feeling "kick-in." The Enneagram focuses more on *why* a person functions in a certain way. For example, an ISTJ may be a Loyalist, a Perfectionist. or a Status Seeker. Renee Baron and Elizabeth Wagele, in their book, *Are You My Type? Am I Yours?* (1995), attempt to relate specific MBTI types with particular Enneagram types. MBTI does not hypothesize why an INTJ or ESFP turned out as they did. Might it be indeed possible that certain types of deprivations initiated not only particular emotional orientations as in the Enneagram, but also certain orders of preference among Jungian functions as in MBTI?

Also, while proponents of MBTI identify a preferred and weaker side on each of the four dimensions and suggest that one indeed try to strengthen somewhat the weaker sides, the MBTI system does not propose specific growth paths. In theory, a majority of people will not change MBTI types: one will not transverse all four temperaments in order to become more integrated. In fact, attempting to be equally balanced on all four functions is discouraged in MBTI literature. However, the Enneagram, through its directions for integration, explicitly encourages a movement through other types for personal growth, and greater self-awareness and balance. A person of a specific Enneagram type will theoretically remain of the same type. However, by passing through the other types on her growth path, the person will become a "redeemed" version of her original type.

Both the Enneagram and the Myers-Briggs type systems can contribute significant insights for the IT professional who is trying to become more integrated and more aware.

APPLICATIONS IN IT

Having been introduced to the basic benefits and drawbacks of applying the Enneagram, an IT professional, particularly a manager, may be wondering when, how, and to what degree ought an IT department to get involved with Enneagram typing. A primary obstacle is the lack of a uniformly recognized instrument that would assign a person to a type. The "technical" mentality of IT may balk at an approach that encourages individuals to study, think, feel, and "see which one fits best."

Secondly, there must be a specific, identifiable benefit in view. The Enneagram offers a deeper awareness of one's underlying emotion(s) and one's self-defeating tendency while under stress. Furthermore, it offers directives for a conscious re-direction of one's inner energies to provide more inner freedom and harnessable mental power. The benefits are there.

Another question for an IT manager may be whether the organization should get involved formally, or simply encourage employees to examine this as a possible source of self-improvement. As well, one may need to consider whether using the Enneagram alongside Myers-Briggs may appear to be "overkill."

For organizations willing to address personality issues among IT workers, it is likely that the Myers-Briggs Indicator will indeed be the first choice. It is more well-known, has one accepted classifying instrument, and has been applied to IS in a noteworthy way. It is suggested here that it is probably a good idea to let employees first be quite familiar with MBTI and its various possible applications in their work. This may take from one to two years. If a number of employees seem to have "warmed up" to MBTI and have indeed recognized at least some specific work benefits, they may be ready to examine the Enneagram.

A good way to begin may be to invite a qualified Enneagram presenter (preferably one with some exposure to IT work) to give a one-day workshop to those interested. It would be important to have noted the similarities, differences, and the complementary nature of MBTI and the Enneagram. After such an introductory workshop, the organization may either leave further work entirely up to each employee or consciously promote further awareness and discussion (through informal monthly meetings, newsletters, etc.). In any case, it would be very desirable for "success stories" to be publicized within the organization.

Particular areas of application for the Enneagram would be individual stress management and cooperation among collaborators. Differing reactions to specific, challenging, and recurring situations (e.g., changes in user requirements) can be noted and brought to light. Furthermore, Goldberg suggests that even different organizations (e.g., the government, the army, a research department, a production line) may be identified with different Enneagram types. This prompts one to consider whether different IT organizations can themselves exhibit traits of one type or another. Alternatively, different subdivisions of an IT organization may be "framed" in particular mindsets and thus prone to specific disintegrative attitudes.

In any case, the best catalyst for promotion of the Enneagram within IT is the presence of significant intrapersonal as well as interpersonal results. Reduction of debilitating attitudes and behaviors, increased consideration of other viewpoints, higher tolerance of ambiguity, and greater freedom for creativity and change are all very welcome realities in today's IT environment. Can the Enneagram, indeed, initiate significant improvement?

FURTHER RESEARCH AND DEVELOPMENT

With increased applications of the Enneagram to work situations in general, MIS researchers may be wise in examining more seriously ways in which this personality system can assist in the field of information systems and technology.

Following are potential areas for investigation:

1. How can Enneagram types in IT be identified in a more well-defined manner?
2. What is the distribution of Enneagram types within IT? Does this appear to differ in different types of IT organizations?
3. What types seem to be attracted to which positions within it?
4. How do the different types react to typical stresses within IT? Are reactions consistent with Enneagram theory?
5. In situations where employees did not fit well in certain IT positions but subsequently perform quite satisfactorily in other IT positions, can the lack of fit be traced to specific Enneagram traits?
6. What are reactions of different types in IT to change? Are reactions predictable?
7. What are the effects of some types on others on IS development teams? Are there common, widespread emotions and behaviors that can be modified with Enneagram consciousness?
8. Do there indeed seem to be different IT departments with specific "IT department type"? Or, are there specific sections of an IT organization that tend to demonstrate a certain type (e.g., a Two help desk; a One quality assurance team)?
9. How do different types of IS analysts interact with different types of users?
10. How does Enneagram type influence the leadership style of IT managers at various levels, and how do subordinates of different types relate to particular leadership styles?
11. What is the relationship between MBTI type and Enneagram type among IT employees?

The major areas of Enneagram influence will hopefully be management of stressful situations (unrealistic deadlines, unexpected requirements changes), management of change (new technologies and methodologies, new application domains), and synergy in teamwork. Such areas may warrant significant interdisciplinary research.

There are also not only research, but development efforts that can be made in attempting to enhance psychological awareness within IT. Firstly, can a type-identifying questionnaire be developed with content related to both personal and IT work issues? What degree of "validity" would such a questionnaire demonstrate? Secondly, with enough empirical data on hand, a booklet can be produced specifically addressing each type within IT. Modes of operation, attitudes, reactions, and common feelings can be pointed out. Directions of integration and disintegration, according to the Enneagram, can be explained. References to general Enneagram books, as well as books applying the types to work, can be provided. Such a "booklet" can well take the form of a Web-based document. In time, such a site can contain video clips of each of the types and even scenarios, enacted from actual work situations, depicting visually how each type reacts to such situations. Team psycho-dynamics can be illustrated on video clips and related to the Enneagram.

At conferences attracting larger numbers of IT professionals, nine smaller rooms can be reserved for a given time, so that people of each type can mix with others of the same type in one room. IT people who are just beginning their Enneagram consciousness may visit two or three rooms that most likely reflect their type. Such interactions may assist them in zeroing in on their likely type.

Through a central website for IT Emotional Intelligence, nine chat rooms can provide a forum for fruitful discussion, and related bulletin boards can provide stories of increased professional effectiveness and personal growth through application of the Enneagram. Such feedback may be very valuable for initiating new, interdisciplinary research efforts that can be spearheaded by MIS academics.

A noteworthy, useful potential item for development might be a video on how to recognize types in the course of a typical day's work for various categories of IT employees. Then, viewers can be shown what can be gained from such recognition in terms of freed-up inner energy, and ultimately increased effectiveness, synergy, and productivity.

Projects and initiatives (some possibly initiated by enthusiastic Sevens) will lead to other developments. However, deeper awareness and widespread communication are essential. With the IT professionals' high need for growth, this area can indeed become a model for other professions regarding impact of personality consciousness.

FOR THE BEGINNER

You, as an IT professional, may have started your "personality literacy" with Myers-Briggs. If so, you may likely have found it structured, well-documented, and credible. Now, you are facing the Enneagram. A first question that may arise relates to its credibility. Although the diagram is indeed a structure, the way of identifying one's type can seem very arbitrary. Thus, after an initial familiarization, at least to the degree outlined earlier in this chapter, it may be useful to identify among the people in your life who appear to be "typical" Enneagram types. For example, an aunt may be a typical One, a nephew a definite Five and a subordinate, a very probable Seven. Interact with these people with Enneagram material in mind, see what motivates them, note how they might react under stress.

A first effective way of establishing personal conviction about the Enneagram can indeed be people-watching. In time, you may notice striking parallels between specific people and Enneagram type descriptions, for at least four or five types. You may then start to believe that there "really may be something to this."

Secondly, you will likely observe yourself. If you cannot identify with a type immediately, you will likely eliminate at least four or five types as not applicable. You may then take an Enneagram indicator test, available on the Web. Remember, though, that this is an *indicator* only. Look at the top two or three candidates for your type. Reread the type descriptions. Perhaps you would like to refer to several other introductory books on the Enneagram; each one will provide additional insight. Alternatively, you may wish to enroll in a short workshop on Enneagram basics. If, at this point, you are satisfied that you have found your type, observe yourself in professional and personal environments to corroborate this. If you are still in doubt as to your Enneagram identity, try to narrow the choices to two, even if it takes some time and self-observation to do so. Once you are satisfied with the top two choices, ask yourself what the key differences are between the two types. List them, point by point. This can help you to zero in. If, however, you seem to be "stuck" at two (or maybe even three) competing types, do not be concerned. Even with this amount of awareness, you can put the Enneagram to beneficial use.

Once you are at or close to your type, observe yourself in various situations. Many valuable insights will be gained. Talk to others who claim to be of the same type; to what degree are you similar? Note especially your underlying emotion—when you feel most "OK"—and your basic compulsion. This, however, will require a deeper level of honesty and authenticity with self.

Assuming you are at a point where you have indeed discovered your likely type and have accepted as normal that type's shortcomings, you probably need to apply this awareness, quite systematically, to your work in the IT area. Note, first, what areas of your IT work are emotionally difficult. This may be dealing with specific users, adapting to change, handling unexpected situations, reaction to criticism, working with tight deadlines, or correcting subordinates. You may have already considered such factors if you have worked with the Myers-Briggs Indicator. Ask yourself: *Why* are these situations difficult to handle? *Why* do they drain your energy? What might be the relationship of your discomfort to your underlying emotion and belief?

Secondly, look at the indicated direction of disintegration for your type. Do you notice yourself going in such a direction when emotionally difficult situations arise? Be as specific as possible. Can you see how going inwardly in such a direction increases your stress level, and decreases productivity and effectiveness?

At this point, it is indeed important to believe (sometimes strictly "on faith") that there is more to you than your current mode of operation. You are more, inwardly, than what you are working with now. Hence the motivation to change. To proceed in the direction of integration indicated for your type does not mean negation of who you are, or a "rebuilding from scratch" of a "new you." It is simply a matter of personal growth, of the discovery of new potentials, already present but not yet actualized. In Enneagram, a Six type, for example, remains a Six, but moves from an "average" or even "unhealthy" Six to a balanced, integrated, "redeemed" Six. This is done by following the steps of integration prescribed by the diagram. After considerable growth, the compulsions are turned to virtues (there is one specific virtue for each type).

It must be said that for some, moving in the direction of integration may be easier than for others. Compulsions are not easy to overcome. As mentioned earlier, willpower may not suffice to move to a different mode; support of others and "a process of inner surrender" may become necessary. With this in mind, persons may not wish to restrict their emotional growth to the Enneagram alone, but may combine basic insights from this system with other development approaches (as, for example, in Chapter IV).

It is important and encouraging to note significant changes for the better, especially if such changes tend to recur. An IT professional who is serious about psychological growth would do well to keep a "log book" on how she has found previously difficult situations much less problematic as a result of a shift in inner orientation. One should be quite specific, noting inner changes in energy and feelings, as well as outer changes in behavior and possibly productivity. Significant feedback could be provided on a website for other IT professionals—for insight, encouragement, and the building of a sense of solidarity. Once it becomes "in vogue" to discuss emotional/psychological growth within IT, particularly among systems developers whose success depends significantly upon interpersonal interaction, Web-based bulletin boards and chat groups could indeed contribute in promoting successful Enneagram applications.

In time, successful applications, if appropriately catalogued, can become a rich source of information for applied researchers. In fact it may well be that interdisciplinary researchers may become motivated to consider seriously research into Enneagram and IT only after observing larger numbers of seemingly successful applications "in the field." *Are there certain predominant recurring feelings in specific IT positions?* If so, how have various types been able to diffuse the negative impact of such feelings through Enneagram consciousness?

As a beginner in applying this awareness to IT work, it is important for you *not to overanalyze*. Personal growth is not an exact science and the Enneagram is only a model. As you proceed, integrate theory and practice at each step. Try to be able to articulate how your emotional reactions are changing when faced with work difficulties, how you are making the shift towards integration, what the fears and hesitations are in shifting and the impact of your reorientation on your work environment. Also, you can recognize that inner growth rarely proceeds linearly—there are backward loops and setbacks, yet lasting progress is possible and thus desirable.

CONCLUSION - PERSONALITY TYPE

IT, and within it, IS in particular, is maturing as a socio-technical field. "Soft skills" are getting increased attention among systems professionals. Conference tracks or even themes for entire conferences are being devoted to related issues. One's personality ("a complex set of relatively stable behavioral and emotional characteristics") can indeed play a significant role in how one reacts inwardly to professional challenges and how one relates to other key players in the IS or user community. Personality awareness, a specific skill, must be followed by capacity to appropriate intrapersonal and interpersonal behavioral changes, another noteworthy ability.

Thus far, publication on actual work relating personality to IS has been dispersed; material on hypothesized IS contributions related to specific types has been indeed rare. This has made an integrated initial understanding difficult to achieve for the already overworked information professional.

These two chapters have introduced the reader to two of the most common personality classification systems, the Myers-Briggs and Enneagram types. Specific connections to IT work were made, noting significant research and application. Practitioners as well as researchers are encouraged to explore new frontiers.

It is hoped that this section indeed initiates personality consciousness in IT on a widespread basis. If such emerging consciousness is followed by extensive communication of successes, perceived failures, and new questions relating to the influence of personality awareness, the "soft side" of IT will have gained a greater degree of prominence, recognized validity, and potential impact.

REFERENCES

Baron, R., & Wagele, E. (1995). *Are you my type, am I yours?* San Francisco, CA: Harper.

Brady, L. (1994). *Beginning your Enneagram journey.* Allen, TX: Tabor Publishing.

Condon, T. (1997). *The dynamic Enneagram.* Metamorphous.

Goldberg, M.J. (1996). *Getting your boss's number.* San Francisco, CA: HarperBusiness.

Palmer, H., & Brown, P. (1998). *The Enneagram advantage*: *Using the nine personality types at work.* Harmony Books.

Riso, D.R., & Hudson, R. (1990). *The wisdom of the Enneagram.* Bantam.

Rohr, R., & Ebert, A. (1990). *Discovering the Enneagram.* Crossroad.

Wagner, J. (1983). Reliability and validity study of a Sufi personality typology: The Enneagram. *Journal of Clinical Psychology, 39,* 712-717.

Wagner, J. (1981). *A descriptive, reliability, and validity study of the Enneagram personality typology.* PhD Dissertation, Loyola University, Chicago, USA.

Chapter III

Cognition, Creativity, and Learning

INTRODUCTION

While personality relates to one's behavior as a whole, *cognitive function* relates more explicitly to mental information processing. Since the majority of system development work and IT work in general involves intellectual functioning, it is not difficult to see that how a person performs "mind work" is a relevant psychological factor in IT work. In fact, it is in the area of cognition that a majority of psychological research in computing/information systems has been carried out.

This section, however, does not aim primarily to present and discuss specific research findings. Rather, by reference to numerous, relevant sources, it aims to present in "layman's terms" main points on cognitive, creativity, and learning styles to a broad audience of IT professionals and academics. It then tries to relate differences in style to effectiveness of IT work, and argues for the benefits of style awareness and conscious cooperation among IT professionals. Finally the inclusion of "style wisdom" in the psychological intelligence set of the IT professional is promoted.

COGNITIVE STYLE

According to Hayes and Allinson (1998), cognitive style is "a person's preferred way of gathering, processing, and evaluating information." Streufert and Nogami (1989) identify cognitive style as a pervasive personality variable. It influences what information in one's environment a person focuses on and how he/she interprets this information.

One main way of dichotomizing cognitive functioning is the *analytic, sequential* versus *intuitive, wholistic* functioning. Some psychologists have referred to the former as "left-brain thinking" and the latter "right-brain thinking" (although other scholars may consider this an oversimplification). The former focuses on "trees," and the latter sees the "forest" in solving problems and coming to conclusions. There are suggestions that link MBTI temperaments with the two predominant cognitive styles. Huitt (1992) identifies NTs and SJs as more linear and serial, and NFs and SPs as more wholistic and intuitive.

We have all likely seen classic examples of each style, either at work or elsewhere. The analytical person abstracts, analyzes, structures, organizes, and plans systematically. He can articulate clearly and may focus on details. However he can miss "intangible clues" such as facial expressions or other "body language." He may also be criticized for being bureaucratic with limited imagination.

The intuitive person integrates many perspectives, finds problems and discovers opportunities, and generates new visions. She is sensitive to both logical and emotional issues, viewing them as one. However, she may overlook important details, may not communicate precisely enough, and may put off decisions.

The analytic is like a tax lawyer, the intuitive, like a criminal lawyer; the analytic like an accountant or chemist, the intuitive like an athlete or artist; the analytic prides himself on intellectual rigor, the intuitive on imagination. As with personalities, most people may not fall exactly into two opposing "boxes"; however, there appears to be a definite preference in one direction.

Adaptor-Innovator

One of the main theories on cognitive styles, along with an instrument to evaluate the style, is the Kirton Adaptation/Innovation theory. This theory of cognitive strategy relates to the amount of structure that a person feels appropriate within which to solve a problem or to embark on creativity.

The Adaptor (left-brained) prefers to work within current paradigms, focusing on *doing things better*, while the Innovator (right-brained) prefers to "color outside the lines," constructing new paradigms, focusing on *"doing things differently."* According to psychologist/consultant Michael Kirton (1989), the Adaptor favors precision, reliability, efficiency, prudence, discipline, and conformity. He seeks solutions in "tried and true" ways, seeks improvement and efficiency, and rarely challenges rules. He is seen as sound, dependable, safe, and conforming and is an authority within given structure.

The Innovator, on the other hand, cuts across and often invents new paradigms. He is more interdisciplinary, approaches tasks from unsuspected angles, and often treats accepted means with little regard. He tends to take control in unstructured situations, but is usually capable of detailed routine work for only short bursts of time. While an Adaptor has higher self-doubt and is vulnerable to social pressure and authority, an Innovator does not need consensus to maintain confidence in the face of opposition.

To distinguish Adaptors from Innovators, Kirton developed the Kirton Adaptation-Innovation Inventory (KAI), a questionnaire consisting of 32 items, each scored from 1 to 5. KAI scores thus range from 32 for a pure Adaptor to 160 for a pure Innovator. The stability of the KAI has been observed over time and culture; it may be measured adequately even early in a person's life. KAI distributions for general population samples from six countries conformed almost exactly to the normal curve with a mean of just above 94 (where the theoretical mean would be 96) (Kirton, 1989). KAI scores did not correlate significantly with IQ, achievement, or creativity tests. This emphasizes that KAI is a measure of cognitive/creativity *style* rather than *capacity.*

Cognitive Style in IT

It is not difficult to realize that there will be both Adaptors and Innovators in the IT profession. It is also easy to see that each style, if properly harnessed and managed, will provide significant contributions to the development and implementation of information systems, particularly Web-based multimedia applications.

The late researcher Dan Couger (1997) applied KAI to a representative sample of IS professionals and found that most were inclined to the Adaptive rather than the Innovative style. This would agree with the preponderance of SJ temperaments in IS. In this situation, however, many IS organizations will develop Adaptor managers who prefer the Adaptor (analytic) style. Valuable contributions from talented Innovators may be ignored or discouraged. This may indeed be dangerous in an area where paradigm changes are paramount.

Lindsay (1985) reports on a consulting case where an international oil company was experiencing a conflict between two department managers and their senior analyst. The managers required that the analyst's work tie in directly to the business plan and not deviate from the established "business climate"—he had to establish credibility with other managers. The analyst accused the managers of being short-sighted and not perceiving the future patterns of business nor preparing adequately for this. When the three took the KAI, the managers averaged 86 and the analyst 117. Eventually, the analyst decided to seek employment elsewhere (in Kirton, 1989). Anecdotal reports suggest that the difference of more than one standard deviation in KAI scores may cause significant problems in employee communication. Again, for the IS professional, the key is awareness.

In IS development, there are tasks that require applying established technologies to new situations (adapting) and others that benefit from entirely new approaches. The latter can usually enhance competitive advantage. However, according to MIS researcher Michael Epstein, "The need to integrate the mechanistic logic of computers with the ambiguous nature of living human systems remain fundamental to the successful design and implementation of computer-based information systems" (1996). Thus, the mature IS developer will increasingly need to combine both views consciously and effectively. Moreover, it is not only a matter of "system" issues requiring an analytic approach and "people" issues requiring the integrative, wholistic perspective. "Right-brain" contribution can indeed be significant in specific areas of analysis, design, and development. Design of user interface and Web pages, as well as multimedia components, clarifying and improving upon user-identified system requirements, process re-engineering for ERP implementation, and identification of domain objects can all benefit significantly from a thinking process that is intuitive and innovative. It remains for interdisciplinary IS research to identify more specifically *how* such "right-brained" efforts complement the traditionally accepted analytic approaches to IS development.

To gain practical insight into the issue of thinking style in IS, one can consider some significant developments in applied computing and hypothesize which style may have contributed most in each case. For example, the concepts of CASE tools or ERP systems may well have resulted from a more traditional, analytic style. CASE mainly automates and integrates existing approaches to system development; thus the CASE concept is an extension of an existing paradigm. Similarly, ERP systems are a large-scale adaptation of previously existing smaller integrated business applications.

However, the OO paradigm is likely the result of the innovative, intuitive thinking style—it models data and related processes in an entirely new way. Such an intuitive style may also be responsible for innovative ways in assisting system users, for example the recording by an EIS of which screens an executive uses most often. Also, considering that system development efforts are often initiated by problems and opportunities, it is quite possible that while "left-brain" thinking can address existing problems adequately, it is the "right-brain" approach that can capitalize on new opportunities from an IS perspective.

RESEARCH FINDINGS

The issue of cognitive style in IS has been one of the more researched psychological factors. We now examine several significant findings and then suggest areas for further investigation. Regarding information processing in general, Churchman and Schainblatt (1965) and then Huysmans (1970) suggested that individuals may ignore information presented in a format that does not fit their cognitive style. Applying this insight to IS, Mason and Mitroff (1973) emphasized the importance for IS developers to consider the user's cognitive style. They advised, "With some imagination and creativity, designers should be able to give each manager the kind of information systems he or she is psychologically attuned to and will use most effectively."

Bariff and Lusk (1977) also advised that "the successful development and implementation of an information system should explicitly involve consideration of the psychological disposition of the system's user." They recommended that designers "match report content, format, and presentation mode with the user's psychological structure." They suggested that the use of psychological tests with recognized validity and reliability to evaluate user cognition represents a step toward a more systematic basis of MIS design.

A strong scholarly contribution during the initial years of connecting cognitive style to IS design was made by Benbasat and Taylor (1978). They stated emphatically, "The human decision-maker must be included in any complete analysis of MIS design since he is the user of the information or output of the system." However, they added, "Cognitive styles represent only some of the characteristics of information system users which may influence MIS design. Characteristics such as intellectual ability, attitudes, demographics (age, education, etc.), and cultural background would also be expected to figure prominently in MIS design." They present Witkin's definition of cognitive styles as "…characteristic modes of functioning that we show throughout our perceptive and intellectual activities in a highly consistent and pervasive way," and consider such styles as "promising variables" which "appear central to MIS design."

Benbasat and Taylor further commented on the lack of a uniform measure or classification for cognitive style. In addition to the analytic-intuitive classification, they present those of cognitive complexity and field depen-dence-independence. The authors point out that cognitive complexity deals with differentiation, the number of dimensions extracted from the data, ar-ticulation-fineness of discrimination and integration-complexity of the rules used to combine data. This approach to assessing cognitive style relates to both the amount of information decision makers tend to use and the degree of focusing in the use of data they exhibit.

A cognitive style that measures field dependence-independence has been developed by Witkin (1971) and can be assessed by the *Witkin Embed-ded Figures Test*. The central characteristic here is the ability to overcome the influence of an embedding context. The field-dependent does not easily separate parts from their context; he looks at the totality of the situation (e.g., not separating a person's words from her facial expression). The field-independent can separate parts from their context (e.g., may consider words but not body language). The former style parallels with the intuitive and the latter with the analytic.

Looking at cognitive style based on an analytic vs. heuristic (intuitive) approach, Benbasat and Taylor referred to other studies that showed analytic decision makers preferring reports that have formulas embedded in the text and are quantitative. They then identified a challenge for the MIS designer—to find some kind of decision aid to help the problem-solving task of the heuristic decision maker. They also pointed out that heuristic decision makers need to have more data search capabilities prior to reaching decisions. Since they rely on trial and error, a system capability that can show trends and period-by-period comparisons would suit such individuals.

Finally, Benbasat and Taylor suggested that the systems analyst broaden his perspectives to better satisfy the psychological needs of the user-man-agers. They also proposed that the analysts *develop cognitive style profiles of their system users*, either by informal interviews or by using established psychological tests.

An important question that arises here in the mind of the reader of this paper is how one would accommodate different cognitive styles of users of the same system. Options would need to be presented to users. But, how many options, and at what points in the system? Nonetheless, the work above certainly legitimized concerns regarding cognitive style in MIS research.

Davis and Elnicki (1984), in their paper, "User Cognitive Types for Deci-sion Support Systems," emphasized that a decision support system needs to communicate information to a decision maker in a style that is congruent with the decision maker's decision process. They conducted exploratory research experiments on 96 MBA students regarding the use of a system to support production management decisions. They used the Myers-Briggs Indicator to identify four cognitive types: NT (Intuitive Thinker), NF (Intuitive Feeler), ST (Sensing Thinker), and SF (Sensing Feeler). They found that ST managers would improve performance with graphical, raw data reports; SF managers would appreciate tabular raw or tabular summarized data, NT managers would synergize with tabular raw data, while NFs would relate most to graphical, raw data reports. In their experiments, the authors identified three performance measures: time taken to make the decision since receiving the report, user confidence in this decision, as self- reported on a scale of 1-10, and cost incurred as a result of each of the production decisions.

Although their work did show a connection between cognitive style of the user-manager and the type of information best suited for effective semi-structured decisions, the authors categorized their research as "exploratory." They also stated that "no methodology presently exists for implementing system design with cognitive type." Furthermore, in discussing prior research, the authors indicated that "experimental results on the effects of report format and level of data summarization have not been consistent." They pointed out, for example, that Amador (1977) and also Lusk And Kersnick (1979) had found that users with a tabular report format performed significantly better than those with a graphical report. However, Benbasat and Schroeder (1977) found the opposite effect on performance.

A damper on the enthusiasm for cognitive style research was provided by the "sobering" article by Huber (1983), "Cognitive Style as Basis for MIS/DSS Design—Much Ado About Nothing?" Noting inconclusive research results, Huber questioned the need for the analyst to concern himself explicitly with users' cognitive styles. One could introduce enough reporting/usage flexibility in information and decision support systems that users would select preferred ways of using the system without either analyst or user being explicitly aware of cognitive style.

Huber also cautioned that fitting DSS to user styles may actually reinforce cognitive bias in the decision makers. This, however, begs the question, "Do not all decision makers possess some type of cognitive bias"? As well, Huber observed that "the currently available literature on cognitive styles is an unsatisfactory basis for deriving operational guidelines for MIS and DSS designs." At that point, research had been done, but there was little that could appreciably influence the daily work of an IS professional.

While most of the literature cited above applies cognitive style exclusively to the user of information systems, Robey's article (1985) provided additional perspective. While he acknowledged the validity of Huber's concerns, he used those concerns as a catalyst for new insight. Firstly, he addressed the issue of the flexibility of DSS functioning and reporting options. Huber suggested that if DSS were to be made flexible enough, users would adapt this to their preferred way of use without explicit consideration of cognitive style on the part of the user or designer. However, Robey pointed out that in deciding on *possible alternatives* for the flexibility of a system, cognitive considerations could indeed contribute. Responding to Huber's concern of reinforcement of cognitive bias, Robey raised the possibility of using a DSS to *complement* the user's preferred style, if the user could overcome a threshold resistance to using a system to which he is not accustomed. He suggested that managers, perhaps aided by personnel specialists, should conduct an ongoing study of their styles for a more complete awareness of how they make decisions.

In addition, Robey presented one very significant insight: cognitive style considerations will become more central as *we study the system design process itself,* and the styles of not only the users, but also the designers. In developing today's multifaceted, Internet-based business systems, might it not be indeed advantageous to use cognitive style consciousness among developers to enhance synergy and effectiveness on development teams? After Huber, Robey, and others, a question of interest in the late '80s related to seeing whether, in a highly customizable system as to both interfaces and output, user cognitive style needs to be an explicit factor when designing such a system.

Tan and Lo (1991) presented empirical evidence from studying 160 senior administrators and their secretaries in an educational institution. They investigated whether after users customized an adaptable office automation system interface, user cognitive style was a factor in system success as perceived by the user.

The study did not find evidence that the cognitive styles of the users had any significant impact on the success of an OA system that had its user interfaces customized to the user's requirements. Thus the researchers concluded that, if the interface to an OA system can be customized, then system designers do not need to consider users' cognitive styles explicitly when they are designing the system. This is analogous to saying that, if enough food groups were placed on a banquet table, one would not specifically need to canvas the diet preferences of the attendees. However, is it possible that cognitive considerations may indeed assist designers in determining the types of options available for customization?

With somewhat inconclusive and possibly conflicting research results by the beginning of the '90s, we witness yet a new research slant in 1997 in the work of Spence and Tsai (1997). These researchers wondered whether a particular system user's cognitive style will remain the same over a variety of tasks and in a variety of environments (e.g., would the same person exhibit the same style in the task of assessing a room's suitability for a movie scene and assessing the room's need for a restructuring by carpentry, plastering, and painting?). Thus, Spence and Tsai focus their attention on cognitive *processes* rather than on cognitive *style*. A cognitive process relates to the approaches used in sensing, concept formation, comprehension, problem solving, decision making, research, composition, and oral discourse (Marzano et al., 1988). A cognitive process, according to the authors, involves using a sequence of cognitive skills to process information. Thus, the *pattern* of skills characterizes the cognitive processes. The Marzano source identifies 21 cognitive skills, aggregated into the following categories: focusing, information gathering, remembering, organizing, analyzing, generating, integrating, and evaluating. For information processing and decision making, the higher the complexity, the more sub-tasks the decision maker must complete.

Using experiments with 101 volunteers, the researchers proposed that cognitive *style* has no impact on the cognitive *process* used during a decision-making task, whether the task be quantitative, qualitative, or of high or low complexity. Cognitive styles were considered as "analytic" for Myers-Briggs thinking types and as "non-analytic" for feeling types.

Results did not support the impact of cognitive *style* on an individual decision-making *process*. Results suggested that "subjects with a particular cognitive style do not always use the same approach to solve problems." In the experiments, subjects changed their decision-making processes across four experimental tasks. Thus, it seems that the task influenced the processes rather than the cognitive style. Thus, the authors concluded, "Since cognitive style only indicates general traits of information-processing behavior, cognitive style is neither appropriate nor useful in providing operational guidelines for IS design characteristics which are task based."

Clearly, more research along this line appears warranted, with experiments involving different tasks (planning, controlling, scheduling, etc.) and different information systems in various settings. Definitive conclusions certainly appear premature.

Most of the research outlined so far has concerned itself with cognitive style involved in using system output for decision making. The output is usually that from traditional "reporting systems," although it may be presented in different forms and in different degrees of aggregation.

However, the scope of IS is broadening significantly with the availability of enabling technologies. In this context, new questions are being raised regarding cognitive style. Palmquist and Kim (2000) explored cognitive style related to efficient use of the Web. Their exploratory study aimed to investigate the effect of cognitive style and online database search experience on the World Wide Web search performance of 48 undergraduate college students. Cognitive style was classified as field-dependent or field-independent (we recall that the former, intuitive, takes in information while considering the context, while the latter, analytic, is more adept in abstracting the information from its context, e.g., words may be assimilated without absorbing the body language). Earlier studies—e.g., by Korthauer and Koubek (1994) and Ellis, Ford, and Wood (1993)—had applied the FD/FI cognitive construct satisfactorily to hypermedia contexts. Web search performance was defined by: i) time required for retrieving required information, and ii) the number of nodes transversed for retrieving what was required.

Research results showed that FD/FI cognitive style significantly influenced the search performance of *novice* searchers. Field dependents needed to spend a longer time and to visit more nodes than did the field independents. However, the influence of cognitive style was greatly reduced in those searchers who had online database search experience. The researchers suggest that online search experience may help the FDs to overcome the spatial complexities of hypermedia information architecture. They then provide preliminary suggestions for Web-based interface designers: "It seems that the FDs, especially those with little or no experience with online databases, might need special attention from the interface designers or those who train Web users. Interface designers may want to incorporate devices that can help the FDs become better oriented and less likely to get lost. Providing a graphical map of their search progress would be an example."

British researcher Nigel Ford, in his recent (2000) article, "Cognitive Styles and Virtual Environments," points out that "virtual" environments "enable a given information space to be transversed in different ways by different individuals, using different routes and navigation tools. He identifies two main categories of learners—"serialists" who examine one thing at a time, with the overall picture emerging much later, and "holists" who examine interrelationships early, concentrate first on building a broad overview and subsequently fill in the details. Ford presents a variety of research results on cognition, learning, and thinking, and then outlines applications to hypertext navigation and database searching (e.g., CD-ROM). He then calls for more research in order to assess the extent to which awareness of global (holist) and analytic (serialist) styles of information processing may be useful in the development of virtual environments. He notes evidence that "when information is presented in a way that matches or mismatches individuals' preferred holist or serialist biases, learning is significantly enhanced or disrupted respectively."

Ford then promotes "intelligent, adaptive" systems in a virtual environment. "Such systems could: i) identify particular learning strategies being used by individual users, ii) classify them in terms of a provisional model of learning, iii) offer individualized strategic support based on the model, and iv) use feedback data progressively to improve the model." He finally states, "The capability of virtual environments to integrate and allow the explicit manipulation of global and analytic aspects of a given body of information would seem to map well onto learning requirements suggested by research into cognitive styles."

Although certainly not exhaustive, the above outlined research provides a representative orientation to the consideration of cognitive styles in the area of information systems. Mostly, the point of involvement has been user cognition related to systems output. Relatively little effort, however, has been expended on studying the role of *cognitive styles of system developers* on a team or of systems analysts during requirements determination. Only most recently have efforts been made to relate cognitive differences to user interfaces, largely in a Web context.

A considerably different area relating cognition to computer-based systems is that of *human-computer interaction.* Here areas of concern from cognitive psychology include thinking (e.g., "mental models"), memory (and its limitations), attention skill acquisition, and language. Appreciable applications have been evident in user interface design. For example, mental models (internal human representations for objects, events, and ideas) help people to deal with the outside world in daily life. Yet, such models are incomplete and often inaccurate. System users also develop mental models for computer-based systems. It is believed that the user interface, in fact, is the basis for whatever mental model of a system a user develops. The more accurate are the user mental models, the more effective is the use of the system.

Human-computer interaction has likely been the most direct link between psychology and information systems. For example, Microsoft employs cognitive psychologists to contribute to interface design. Such a psychologist had been working on design of help systems for the next version of MS Office.

She provided valuable suggestions for creative, yet also reliable and efficient test designs.

However, human-computer interaction study is effectively limited to human cognitive psychological characteristics of system users rather than developers. It is also not as directly concerned with differences in cognitive style. The main psychological effects on a person's interaction with a computer system have largely been incorporated in material on user interface design. Since the aim of this book is to propose a significantly broader involvement of psychological factors in the work of IS professionals (with the direct focus on users being somewhat secondary), the topic of human-computer interaction is not examined beyond this recognition. The interested reader may wish to consult the book by Gardiner and Christie (Ed.) (1987) or to refer to the established, authoritative source by Shneiderman (1997).

Possible Future Directions

In the research world, consideration of cognitive style in IS has produced somewhat inconclusive results. The profession, as such, has not embraced this concept to any appreciable degree. Perhaps, indeed, this topic is too localized to be the starting point for a deeper psychological awareness that could improve significantly the work of an IS professional. Yet, when considering cognitive style as part of a psychological "package" that includes personality type, creativity, communication, and basic life vision, the motivation will be there for system professionals to give this factor its rightful due. Should this happen, the interdisciplinary research community is likely to take note and expand established frontiers.

In 1997 Steinberg declared, in the *American Psychologist*, that cognitive style considerations are indeed "in style." A prevalent theme in cognitive style, as we have seen, is the partitioning into an analytic and a "wholistic" (intuitive, heuristic) approach. Such differences no doubt exist among system users as well as system developers. With the scope of information systems continually and rapidly expanding, the many problems to be solved and decisions to be made will no doubt benefit from a functional, conscious synergy among workers with differing styles. Huitt (1992) stated a researchers' opinion that "the problem-solving techniques…are most powerful when combined to activate both the logical/rational and intuitive/creative parts of the brain." Agor (1991) stated emphatically that "it is of great concern today that organizations often thwart, block, or drive out intuitive talent," and also "organizations typically know their personnel by formal job title, responsibility, or years of experience—seldom do they know which brain skills are possessed that can be applied to difficult problems." Undoubtedly the latter comment applies quite directly to most IT organizations.

As pointed out earlier, Epstein (1996) stated unquestioningly, "The need to integrate the mechanistic logic of computers with the ambiguous nature of living human systems remains fundamental to the successful design and implementation of computer-based information systems." He further cautioned that "design strategies which over-value either perspective at the expense of the other are likely to be less powerful and more highly at risk than those which incorporate both," and advocated "cultivating conscious awareness of the designer's…predisposing characteristics" within IT.

It is quite probable that the IT profession, and system development in particular, is ready to begin such "conscious awareness." However, an individual system developer needs to be sufficiently motivated and sufficiently supported to achieve this goal. Specific training materials, video clips, etc. can be developed competently and creatively to address cognitive style considerations across the life cycle stages (e.g., in planning, requirements determination, modeling, design, prototyping, and user support). Cognitive style, along with personality type awareness, can indeed become more than a passing conversation topic in IS, especially once many incidents (although initially anecdotal) of improved performance, decreased cost, and increased morale are credibly reported.

FOR THE BEGINNER

Once cognitive style is recognized as a legitimate psychological factor relevant to IS work, the natural question arises how such consciousness can be used within the profession. The first step is for large numbers of IS professionals to begin to recognize their preferred style, quite possibly alongside their personality type. Areas of IS work in which cognitive style can have a significant effect can be identified. Management can then promote a "culture of awareness" that encourages open communication on effective synergy in different thinking styles. Once an IS developer becomes conscious of his strengths and likely "blind spots" related to his preferred cognitive functioning, he may actually seek input from a co-worker with a complementary style. Such an attitude can indeed reflect "professional wisdom" or "emotional intelligence."

The issue of cognitive style is likely to be viewed by many IS professionals as the most "scientific" of the topics presented so far. Since so much of IS work involves thinking and learning, many people would likely not object to finding out more about *how* they think. Thus, most objections from IS workers would not come on philosophical grounds, but perhaps on grounds of a general uneasiness regarding self-examination.

If you, as an IS professional, have recognized potential value in understanding your cognitive style and how it relates to various aspects of your work, there are steps you can take. If you have already completed the Myers-Briggs Indicator, you may wish to start there. Note that some authors have considered the Sensing preference to correspond to an analytic style and an Intuitive preference to correspond to the "wholistic" style.

In research presented earlier, four distinct styles were extracted through the MBTI: NF, NT, SF, ST. This may be a starting point. If you have access to a practising psychologist, you may wish to complete the Creative Styles Inventory (Mc Ber). This test presents the client with 32 choices intended to measure a strong or slight preference for the sequential vs. the wholistic approach. This results in a scale from -64 (highly analytic) to +64 (highly wholistic). You may also wish to take the Witkin Embedded Figures Test, which discerns between field-dependence (wholistic) and field-independence (analytic).

Initially, however, you may simply wish to observe yourself in daily work activities. Do you tend to look at the "big picture" first, concerning a new system: the organizational impact, the functions served, the "players" involved? Or do you mostly focus on the "individual trees" such as data entities, processes, program logic, etc., with all else in the background? When speaking to users, for example, about system requirements, do you absorb mainly factual information, or do you naturally blend this with perceptions about tone of voice, eye contact, body language, etc.? Do you have an easier time with linear step-by-step thinking than with an integrative approach?

You may have an initial assessment regarding your cognitive style. Over time, you may like to keep on re-assessing your initial impression. Once you are seemingly more convinced of your analytic or wholistic style, you may wish to note in which IS tasks your style is a true strength and which tasks are more difficult to carry out with your preferred approach. For example, a programmer/analyst might say, "I feel very strong and competent in designing a processing algorithm and converting it to code, but I feel less at ease in requirements meetings when different categories of users are all pushing for 'their own thing' in the new, proposed system."

Once you have a more thorough, personal self-assessment of your cognitive style at work (with adequate, specific examples), you may rightfully ask, "What now?" To start on a positive note, find many examples of your superior performance which you can attribute at least in part to the strength of your cognitive style. Understanding more clearly and more directly why your style was indeed a strength will enable you to be more cognition-literate and eventually more inclined towards further self-development.

After an initial period of self-appreciation, you may wish to see areas where you tend to struggle more (for many STJ types in IS, this will likely be the less linear, less "structurable" aspects of IS work). Can you see, perhaps, that some of your struggle may be resulting from trying to apply a cognitive style to a problem or situation for which this style is not optimal (fitting the "square peg into the round hole")? What key items of information might you be disregarding (e.g., political consideration in the organization, apprehension among users)? And finally, you can try to assess, as realistically as possible, the degree to which you are proficient in the style opposite to your preference. Do you feel an inclination to investigate how you can develop further your "weaker side"? To help you, the IS profession could develop a very useful resource in the form of an "IS Cognition Website."

It may not be most interesting to pursue such self-assessment alone. Where there is freedom and openness to self-development among co-workers as well as higher management, an IS professional may be well advised to share his/her self-assessment with others. This may allow for valuable feedback, greater trust, and mutual support. In some organizations "special interest groups" on psychological factors in IS might be formed, with guest speakers, reference lists, and possibly an electronic newsletter. The goal can be to become more self-aware and more inwardly developed so as to create better team synergy, work more effectively, and to relate to the users more responsively.

Emotional Awareness

It may indeed be useful for the self-developing IS professional to see how his/her cognitive style literacy is influencing his/her emotional awareness. What influential emotions may arise in the course of your daily work that seems to be quite directly connected with your cognitive style? For example, how might your work experiences with your cognitive style relate to your supposed "underlying emotion" as assumed by the Enneagram? An analytic, structured style, for instance, may relate to control or power, whereas a wholistic style may feel connected, committed, or prophetic as to the competitive advantage of a new system.

Since most IS professionals need to interact with others (some quite substantially), it is likely that your cognitive style will be challenged, often implicitly. It may be a situation of "he's talking apples and I'm asking for oranges." What emotions are common for you in such situations? In what areas of your work might cognitive style differences present noteworthy difficulties? Are such difficulties common for people with your style? Sharing of such insights in a literate, well-documented manner over a widely accessible medium such as the Internet can indeed be a significant step in the psychological maturity of the IS profession.

A Need for Collaboration

It is proposed here that cognitive style (which can be considered an explicit part of personality) is indeed worthy of consideration as a specific psychological factor among most IS professionals. One may not be able to change his essential style, but one can learn to develop "the other side" to some degree, and to collaborate consciously and often openly with the intent of improving communication and understanding.

As long as cognitive style or processes therein remain as the casual research interests of a few, albeit capable academics, the profession will not be impacted significantly. An increasing number of professionals will need to get astutely involved in observation of self and others. Such individuals can begin documenting the positive impact their awareness and growth has been having on their work (in the practice of "action research," as advocated in a later chapter). Professional acceptance and involvement in psychological factors, cognitive style among them, can no doubt motivate more comprehensive academic research and give such research motivation and immediate relevance.

CREATIVITY STYLES IN IT

Creativity can be described as "the imaginatively gifted recombination of known elements into something new," or, according to Poincaré, "a fruitful combining which reveals unsuspected kinship between facts long known but wrongly believed to be strangers to one another." Having examined how a person thinks and perceives, it is in order to consider also *how a person creates*.

A key question of course is whether creativity is truly important for the IT professional of the future. Although certain segments of the IS profession can be seen to be applying the same approaches to slightly different application domains (which, some may argue, is not that creative), there is definitely a subpopulation of IS workers who are charged more explicitly with opening new frontiers and new paradigms.

Thus, some relevant questions in the quest for multidimensional development of the IS professional relate to situations in which a person can indeed create something new in this profession, how people create, and what is needed to motivate and enable people to create. Also, a more specific delineation of *where* creativity is most beneficial within IT can be useful. Robert Glass, in his book, *Software Creativity*, identifies both "intellectual" and "clerical" tasks in software development. Among the intellectual he includes constructing models and representing relationships, as well as identifying the impact of design changes and generating screen mock-ups. Identifying rules violations, maintaining lists of requirements, and storing versions of a design are counted among clerical tasks. A number of the intellectual tasks, he proposes, will "require creativity in their accomplishment." Glass goes on to state that "creativity is an essential part of the work of software development," and that "analysis and design are arguably the most intellectual and creative of the life cycle phases." This begs the question: What kind of increased awareness can the "enlightened" IS professional develop regarding creativity at work, and how might this be done? A question immediately following is how might such "creativity literacy" contribute to one's professional effectiveness. These questions give rise to a proposal for specific creativity considerations to be given to intended systems.

After having accepted a specific definition of creativity as it relates to system development, the profession would be advised to identify specific creativity points (occasions) in the development of computer-based systems. Perhaps, later, the word "points" can take on a more quantitative character, as a system can be assessed on its "creativity points" as it would on function points. A specific description of what makes each point/occasion creative would also be in order. One could then look at common knowledge on cognitive/creativity styles to propose an optimal style for each creativity point. Empirical research could then assess such hypothesizing.

For an elementary overview of *creativity styles*, we can recall, from cognitive style presentation, the basic split between those who are more sequential, analytic, and those more wholistic, heuristic. This characteristic carries over into creativity. *Adaptive* creativity (apparently practiced more by analytics) accepts a current paradigm but extends approaches within it. *Generative* creativity (belonging more to heuristics) challenges the current paradigm and creates an entirely new one (in a magazine profile on a high-performing IBM employee, it is stated her favorite pastime is "coloring outside the lines"). Glass states, "Formal methods, perhaps, are appropriate for solving mechanistic and well-understood problems or parts of problems; heuristics are necessary for more complicated and creative ones." With traditional, textbook initiators of system prospects being identified as problems and opportunities, is it possible that the adaptive creativity style may be more suited to the former and the generative style to the latter?

One can also hypothesize, for example, that the introduction of CASE tools and ERP systems reflect adaptive creativity within IT, while the introduction of the OO paradigm not only to programming but to design and analysis can be considered as a prime example of generative creativity, since an entirely new approach to information modeling and processing is being promoted. The Creative Styles Inventory (CSI), previously mentioned, is a psychological instrument that assigns a score to indicate analytic ("left brain") or heuristic ("right brain") cognitive preference as well as the strength of such a preference. Knowing such information about oneself, and having collaborated it in the course of one's work, could indeed enable a system developer and his/her manager to make optimal use of his creative potential. (In teaching this material in an ISM course to MBA students, I asked people to identify most preferred and least liked courses in the program. Invariably, high analytics preferred finance or accounting, while high heuristics opted for organizational behavior or management strategy.)

Thus, a beginner's orientation to his/her creativity style can well be initiated by an instrument such as the CSI. The person can then list incidents where he/she participated in adaptive or generative creativity in the course of IT work. Where was there relative comfort vs. a somewhat unwelcome challenge? Can one pinpoint specific sources of discomfort, likely from requirements to create outside of one's main style. Can such consciousness be brought to light in the profession in general and on specific projects in particular?

After having imitated a systematic observation of both self and the work environment regarding creativity, the IS professional would be well advised to examine the extensive work of the late Dan Couger. He played a central role in the establishment of The Center for Research and Innovation in Information Systems at the University of Colorado Springs. Also, he had pioneered the systematic application of creativity concerns to the area of information systems.

In his book, *Creativity and Innovation in Information Systems Organizations* (1996), Couger identifies four approaches to creativity. An attempted outline of these approaches is provided:

1. *Modifying:* What can we adapt or improve upon? For example, the microcomputer itself can be seen as a scaled-down adaptation of a large computing machine.
2. *Experimenting:* What existing ideas can we combine and test? For example, the idea of the need to exchange business documents between vendor and customer and the idea of electronic telecommunication between computers can be combined to facilitate Electronic Data Interchange.
3. *Exploring:* What metaphors can we use to change our assumptions? For example, the waterfall as a metaphor for life cycle stages may cause one to question the assumption of a direct sequence of stages without the possibility of backward looping.
4. *Visioning:* What can we (realistically) imagine as the ideal solution over the long term? For example, visioning the ideal of an Executive Information System that integrates internal and external data, provides choice of output format, and enables a drill-down option, can give rise to a new class of systems with their own unique impact.

It may seem that the first two approaches coincide more with the adaptive style of creativity while the last two reflect the generative. Of course, a person is not limited to employing only one approach throughout his career, but preferences are likely to be formed.

A main point in Couger's work, which applies to a large contingent of IS professionals as well as students, is that *creativity can be taught.* Couger states unequivocally: "Everyone possesses creativity; this God-given talent needs to be resurfaced in most individuals." He then identifies common barriers to creativity: perceptual blocks (accepting as fact data that are really unsubstantiated assumptions), intellectual blocks (paradigm fixation, restrictive thinking), cultural blocks (belief in the uselessness of daydreaming), environmental blocks (lack of logistic and psychological support), and significantly, emotional blocks (fear of failure and ridicule, inability to relax, and buried emotional pain).

Couger introduces the reader to specific creativity techniques. For example, in Interrogatories, one asks: who, what, where, when, and how? For instance: How might this system be used for auxiliary purposes in addition to identified, intended uses? Why would we want to provide this service to our clients? Who could best use the system? Where could it be used? The Manipulative Verb Technique uses a list of verbs to manipulate the problem in order to come up with new perspectives, e.g., eliminate, rotate, flatten, add, invert, separate, repeat. For example, can we *divide* users into levels of sophistication and design different learning methods for the different classes?

Couger also proposes creativity techniques for the development stages of requirements determination, logical design, physical design, program design, etc. He presents lists of work tasks for the programmer/analyst, systems manager, and operations personnel which possess a low, medium, or high level of creative opportunity. In addition, he presents compelling examples of significant cost effectiveness of applying creativity to information systems. Couger's work legitimizes a serious interest in creativity issues in IT and points out areas for further investigation. In the same spirit, Bostrom and Nagasundaram (1995) declare that institutionalizing creativity requires much more than exhorting employees to be creative in their work. Such a declaration, in the *Journal of Management Information Systems*, is a definite step towards psychological literacy within IS practice. The authors point out that studying creativity can include the creative person (or group of persons) as well as the created product, the process of creating, and the environment that enabled the creativity. They also differentiate creativity *level* (how creative are you?) and creativity *style* (how are you creative?). Individuals having the same creative level (based on some creativity level test) could have distinctly different creativity styles.

Bostrom and Nagasundaram then allude to Kirton and identify the adaptive and innovative as the two principal preferred creativity styles (here, we see a direct involvement of cognitive style in creativity). They go on to classify the *products* of creativity as Paradigm-Preserving and Paradigm-Modifying, indicating a likely continuum between the two extremes. They then present several creative problem-solving processes such as brainstorming and guided fantasy. The authors point out the difficulty in evaluating the quality of creativity which is "hard to operationalize and tedious to measure." A research environment may reserve an "exceptionally creative" label for a true paradigm change, while an industrial setting may reward applaudingly new cost-cutting approaches that are well within the established paradigms.

When we consider the environment that would enable creativity, we can relate to the research of Amabile and her associates (1992). Factors that stimulated creativity were freedom (control over one's work), trust and communication among the work group, challenge, organizational encouragement/ recognition, and sufficient resources. On the other hand, harsh criticism of new ideas, politics, a risk avoidance atmosphere, and excessive workload were found to impede high creativity. On the same topic, Couger had found that workload and organizational creativity impediments were perceived as being significantly higher in IS than for other professions requiring creativity.

Another factor worth considering is creativity-enabling technologies, particularly for virtual work teams. In the earlier referenced research, Bostrom and Nagasundaram considered the possible contribution in this area by group support systems (GSS) and declared, "GSS are particularly well suited for group creativity tasks." Simultaneous and anonymous input of ideas may allow for more possibilities to be brought forward. However, the researchers raised noteworthy questions such as:

- How does the use of a GSS interact with individual creativity style?
- How does the use of a GSS interact with the creative problem-solving process?
- Is the development of new GSS creativity tools warranted?

As well, we have seen that as far as future research on IT creativity is concerned, it may be imperative to develop more effective criteria to evaluate creativity, criteria that may well differ in different contexts.

FOR THE BEGINNER

The issue of creativity in IT is likely to increase in importance. As individuals within the field become more psychologically aware, there will be many "beginners" among system developers, project managers, and even IS educators who will deem it worthwhile, maybe even necessary, to develop a functional awareness of their own creativity styles and those of their co-workers.

Worthwhile "starters' questions" are:

- How creative am I as a person?
- How creative have I been in my work?
- How creative would I like to be in my work?
- How much would increased creativity on my part increase my job satisfaction?

As a system developer (using the term rather broadly), you may consider yourself indeed creative or hardly at all. You may first need to define clearly how you understand "creativity." Perhaps (as with the majority of IS workers) you are quite comfortable with adaptive creativity, but believe that it is only ground-breaking generative creativity that warrants the label. Being more explicitly aware of your cognitive style may be of assistance in assessing your capacity for true creativity.

Once you understand your preferred creativity style, you may wish to consider other areas of your life where you now recognize your creative efforts. Next, are you aware that creativity can indeed be learned? You may wish to consult the Internet or a local bookstore for an introductory creativity program. For example, the book, *Unleashing the Right Side of the Brain—The LARC Creativity Program*, by Williams and Stockmyer (1987), provides a systematic approach for unlocking creative potential.

If introductory exercises have heightened your curiosity and increased your belief in your creative potential, it may be time to assess more thoroughly creativity in your work. If you have not been that creative at work, is it mostly because of the nature of your tasks, environmental impediments, or personal blockages? If you have been continually creative, it may be in order to examine the emotional results of your creativity efforts. Do you feel a peaceful, balanced fulfillment or an "inordinate high"? Might you run the risk of being *addicted to creativity*, i.e., using it as a mechanism to avoid dealing with other emotional difficulties?

Carrying out such self-analysis may indeed be intimidating. However, if the entire profession is actually moving towards embracing psychological awareness in its various aspects, you may indeed have many work partners with whom to share your concerns, discoveries, and, perhaps, delights.

As a project manager (or even a higher-level IS manager), you no doubt have realized the reality of radical and continuous change in your profession. Having discovered how your personality type and cognitive style have affected your own career, you may now wish to apply this valuable awareness to your subordinates. To amplify your motivation, you may wish to refer to Couger's book, particularly to the actual illustration of startling cost effectiveness and competitive advantage gained through a concerted application of creativity.

You may wish to consider the individual tasks making up a project, noting where and what type of creativity is in order. You may begin to note any significant improvements in product quality, duration of development, increased morale, etc., as a result of an explicit creativity consciousness among your subordinates. If you can cultivate an atmosphere of trust and open communication, valuable insights regarding creativity may surface from the team members. It is important to acknowledge the differences between adaptors and innovators, but even more important to respect each style. This comes through awareness and an adequate degree of self-confidence.

If you are a higher executive, you will realize what an impact increased creativity will make in your particular environment. You may wish to invite a "creativity consultant" to speak to the IS staff. Above all, you will wish to assess honestly any existing, significant organizational impediments to the creativity of your staff.

Creativity in IT need not be limited to systems development, but can extend to related areas such as IT staffing. For example, to cope with IT staffing shortages, organizations have creatively turned to citizens of small rural communities (as an alternative to offshore outsourcing), workers aged over 50, exceptional high school students (initially working part time), and disabled (e.g., visually impaired) workers. Also, an organization has been formed by a Jesuit priest in the U.S. to train disadvantaged young men and women to work in IT, particularly in Web development.

A broad, concerted, and systematic concern with creativity within IT can be greatly facilitated by an informative website and a dynamic virtual community. With such realities, the creativity concerns among MIS academics would undoubtedly broaden. For example, more groundbreaking work may be warranted on criteria used to evaluate creativity, the type of tasks/situations most fitting for each style, specific creativity development exercises within an IS context, and creativity on virtual teams.

The issue of creativity has begun to be addressed more directly within IS education. At one U.S. university, bonus points are offered to students for such creative efforts as suggesting additional relevant business opportunities besides those called for, applying an additional financial criterion for evaluating economic feasibility, or using an alternative representation technique for program specification. An "A" grade is given only to those students who earned a certain minimum number of bonus points for optional items. As well, the process of creative thinking is supported by all available techniques proven effective in other disciplines and applicable to the IS field. Creativity is fostered by organizing the instructional environment with substantial flexibility.

The IS profession cannot ignore the issue of creativity. A system developer's consciousness of his/her preferred creativity style and awareness of when and how that style is particularly impactful will become increasingly desirable. Also, the encouragement and support for IS workers to strengthen and broaden their creative capacities will be increasingly called for by the systems professionals as well as by their management. It will no doubt be easier to address creativity styles if and when the IS profession as a whole embraces a broader psychological awareness as part of its professional development. This will be another step within IT in the development of its professional wisdom.

LEARNING STYLES

Another psychological factor related to cognitive style is *learning style*. The latter can, in fact, be considered a component of the former. Learning in information systems occurs in a number of places. However, not all individuals learn in the same way. Increased awareness of the learning process on the part of the IS professional could indeed contribute to the effectiveness of IS development and use. The most obvious occasion for learning is that of a user needing to learn a newly implemented system, or of a user learning tools to develop his own computer support. However, the system development professional is mandated to learn throughout her daily work. She often needs to appreciate new system development technologies and even paradigms. In addition, she needs to become significantly familiar with the application domain, with the intricacies of the current system, and with the intended requirements for a new system effort.

There have been several attempts to classify learning styles, particularly in the field of educational psychology. McCaulley and Natter (1980) have identified four learning styles based on the Myers-Briggs Type Indicator:

The *abstract-reflective* learner tends to be scholarly and introspective, interested primarily in ideas, theories, and depth of understanding—characterized by introverted intuitive (IN). For this style, knowledge is important for its own sake.

The *abstract-active* learner tends to see possibilities as challenges to make things happen, and likes to explore new patterns and relationships—characterized by extraverted intuition (EN). Here, knowledge is important for innovation.

The *concrete-reflective* learner tends to test ideas carefully to see whether they are supported by facts. He prefers to deal with what is real and factual in a careful, unhurried way—characterized by introverted sensors (IS). For them, knowledge is important to establish truth.

The *concrete-active* learner tends to be active and realistic, and learns best when useful applications are obvious—characterized by extraverted sensors (ES). For them, knowledge is important for its pragmatic value.

We can see the essences of the four styles in: i) theorizing, ii) innovation and exploration, iii) true, factual reality, and iv) active, practical usefulness. We can likely then imagine Style 1 in strategic planning and system modeling, Style 2 in system alternative generation, Style 3 in programming and quality control, and Style 4 in system implementation. Thus, we notice different types of learning present within system development. Awareness of such a reality on the part of developers as well as project managers can indeed be invaluable.

Some research has been reported connecting MBTI, learning, and information systems. Matta and Kern (1991), of the University of Notre Dame, conducted an experiment involving 110 students in an introductory computer course. They taught rudiments of Lotus 1-2-3 to two groups of students—one learned through classroom instruction and the other trough interactive videodisc (computer-aided instruction—CAI). They hypothesized that the effectiveness of CAI is influenced by aspects of the student's personality. Learning effectiveness for both groups was assessed by means of a post-test, identical for both groups.

Results showed that introverts outperformed extraverts in both groups. The sensing individuals performed significantly better than the intuitive for CAI, but not for the classroom group. Of the four MBTI learning styles mentioned above, the INs in the classroom group displayed the highest average; the ISs in the videodisc group were second; the lowest score belonged to the ENs in the videodisc group. It would be interesting to see if results would be different with the software to be learned being GUI-based and not as formulaic.

In systems oriented towards users with differing functions and at different levels, it may indeed be useful for academics and practitioners to collaborate in further applied research as to possibly different ways of training intuitives and sensors.

Chu and Spires (1991), after stating that "working with computers is primarily a cognitive activity," conducted an experiment on 132 first-year MBA students at a U.S. university. They wanted to see if cognitive (learning) style, as reflected by MBTI, was related to computer anxiety. They found that intuitive students had lower anxiety scores than sensors, and thinkers had lower anxiety scores than feelers. They explain these findings by stating that computer work requires manipulation of abstract symbols and requires a logical frame of mind. This research appears exploratory, and it seems indeed premature to expect sensing people to be consistently more computer anxious, especially now, in the GUI age. However, the issue of computer anxiety, particularly among users, does warrant further, systematic attention from both researchers and IS practitioners.

Another well-recognized and much applied classification of learning styles is the one developed by psychologist David Kolb. Kolb's experiential theory of learning proposes that people learn and solve problems by progressing through a four-stage cycle: concrete experience (CE), followed by reflective observation (RO), which then leads to the formation of abstract concepts (AC), and this conceptualization then gives rise to testing of hypotheses through active experimentation (AE). This cycle repeats itself, as shown in Figure 2. In the figure, we see that concrete experience and abstract conceptualization are at opposite ends, and an axis can be drawn between them. The same goes for reflective observation and active experimentation, where another axis can be drawn, perpendicular to the first one. Kolb asserts that learners have a preference for two of the four stages, that is two stages where one immediately follows the other in Figure 2. We can thus say that learners will prefer, in their learning, one stage from each axis. This allows one to identify four learning styles, one for each quadrant in Figure 2.

We can now examine each learning style by going through the quadrants of Figure 2, starting with the upper right.

Figure 2: Kolb's Learning Cycle

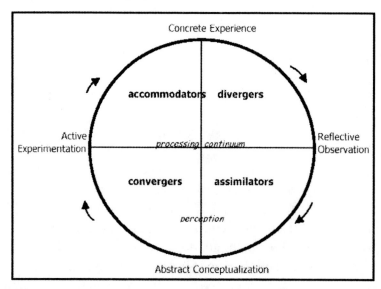

The *Diverger* prefers concrete experience with relative observation. Divergers' greatest strength is in imaginative ability and awareness of meanings and values. They view a concrete situation from many perspectives. The name "diverger" results from the particular ability to generate alternative ideas such as in "brainstorming."

The *Assimilator* prefers reflective observation, then abstract conceptualization. The greatest strength of this style is inductive reasoning and the ability to create theoretical models. Individuals with this style assimilate disparate observations into an integrated explanation. They focus more on abstract concepts than on people, and judge ideas by sound logic and precision rather than practicality.

The *Converger* prefers to employ abstract conceptualizations in order to carry out active experimentation. Convergers' greatest strength lies in problem solving and practical application of ideas. People with this style do best in situations where there is one correct or at least preferred answer, to which they "converge." Usually, convergers are rather controlled in their expression of emotion and prefer dealing with technical tasks.

The *Accommodator* emphasizes concrete experience and active experimentation. Accommodators' greatest strength lies in doing things. They seek opportunities and take risks. Where theories don't fit the facts, accommodators will likely discard the theory. They are called as such since they accommodatingly adapt themselves to changing immediate circumstances. People with an accommodating style tend to solve problems in an intuitive, trial-and-error way.

In order to assess a person's learning style according to his theory of experiential learning highlighted above, Kolb had developed and later improved the *Kolb Learning Style Inventory (KLSI)*. This forced-choice instrument indicates in which quadrant the majority of an individual's learning mechanism lies. Taking the Kolb Learning Style Inventory can indeed offer a person valuable and useful insight into his/her approaches to learning, work, and life behaviors.

Among undergraduate U.S. college majors, psychology, English, and political science majors were mainly divergers; mathematics, chemistry, and physics students were assimilators; engineering and nursing students, convergers; and business students, accommodators. When applying this model to various professional groups, accountants, engineers, and medical doctors were convergers; elementary school teachers, occupational and physical therapists, as well as dietitians and educational administrators, were accommodators; social workers were divergers. These come from various research sources presented in the authoritative book by Kolb (1984).

Learning Styles and Personality

Some investigation has been carried out to connect Kolb's learning styles with MBTI. Introversion and intuition have been linked with Assimilation, thinking and judgment with Convergence, sensing and perception with Accommodation, and feeling with Divergence. Other attempts have suggested extraverted sensing links with Accommodation, introverted feeling with Divergence, introverted intuition (again) with Assimilation, and extraverted thinking with Convergence. Yet, in these attempts, there appear to be inconsistencies and questions. For example, Keirsey and Bates describe an INTP as an abstract theoretician (e.g., in physics, mathematics), yet the INTJ is more of an applied, engineering type. In KLSI, then, INTP should fit with Assimilator and INTJ (likely among others) with Converger. A connection of the MBTI-related styles of McCaulley and Natter with KLSI styles would likely result in the following correspondence:

IN—Assimilator
EN—Diverger
IS—Converger
ES—Accommodator

Table 14: KLSI and the Enneagram

Enneagram Type	KLSI Style(s)
1- Perfectionist	Converger
2- Helper	Accommodator
3- Status Seeker	Accommodator
4- Artist	Diverger
5- Knowledge seeker	Assimilator
6- Loyalist	Converger
7- Fun Seeker	Accommodator, Diverger
8- Power Seeker	Accommodator, Diverger
9- Peace Maker	Diverger, Assimilator

However, here again, for example, the INTP/INTJ situation would not connect with two different styles.

There has not been a widely publicized effort to link KLSI with Enneagram scores, perhaps since, as mentioned earlier, determination of Enneagram types remains more descriptive than psychometric. Preliminary suppositions could be attempted through the attempted links between Enneagram and MBTI, as seen in Table 14.

Realizing that it is hypothesized that each of the nine Enneagram types has an underlying emotion, it may be of interest to see *if specific underlying emotions play any direct role in influencing learning style.*

Learning Styles and IS Tasks

Even at a preliminary glance, it is not too difficult, with the above information, to make preliminary, but at least somewhat well-founded assumptions, about relative strengths of different KLSI styles in various IS tasks.

- *Prototyping, user interface development:* The Diverger can use imagination, assess alternatives, and be sensitive to users; the Accommodator can focus on the practicality and usefulness of the interfaces.
- *Data/process/object modeling:* This is the natural domain of the Assimilator who goes from observation to abstraction (modeling).
- *Programming/testing/quality control:* The Converger takes abstract models and actually experiments by developing and testing code; we recall that this style does best when there is a single correct answer to a problem.
- *System implementation:* The Accommodator is the doer and can be flexible towards sudden changes—his style can be very useful here; the Converger may also fit, preferring a specific outcome after active experimentation
- *Telecommunication:* In network design, the Converger could be competently assisted by an Assimilator, especially in complex, wide-area situations; generally this is likely the domain of Convergers.
- *Business Process Re-engineering (BPR):* In the earlier stages of modeling, alternative ways for an organization to operate so as to take maximum advantage of a major system change can be proposed by Assimilators, possibly aided by Divergers (if both are conscious of each others' differing styles, this could provide very desirable synergy); in the latter stages of BPR when the new system is implemented and new processes must be taught and initiated, Convergers and Accommodators can rise

to the occasion (here we see how in large system adoptions, e.g., ERP systems, specific learning/work styles or preferences, if judiciously considered, can indeed strengthen the implementation process).

- *Website design:* The parts that the users see, particularly if involving multimedia, would be a natural concern for Divergers, whereas the technical underpinnings could involve Convergers, possibly assisted by Accommodators.

- *Help desk/EUC support:* This area undoubtedly belongs to the Accommodator style, which prides itself in adapting to changing circumstances ("fire-fighting").

We have seen a rudimentary attempt to match KLSI styles with major IS tasks. The reader may, in fact, wish to estimate temporarily his/her KLSI style and comment on the above suggestions. Further academic research should indeed cast more justified light on the application of Kolb's learning styles to system development and implementation. At such a point, well-known formal approaches/methodologies, such as the Unified Modeling Language (UML) for object-oriented systems, may specifically incorporate the learning style factors. This would indeed reflect a maturation and evolution of the IS profession.

As we have seen, KLSI learning style awareness can indeed have an influence in the systems development process itself, and not only in the training of end-users in the workings of the new system (although the latter context might be of most immediate association of IT with the phrase "learning styles").

RELEVANT RESEARCH

Some noteworthy research findings have emerged connecting Kolb's learning styles with IS, but they have related mostly to the training of end-users.

Sein and Robey (1991) administered the KLSI to 80 novice computer users, who were trained to use an e-mail filing system by two different methods. The first method used an analogical model of a filing cabinet, while the second method used abstract schematic diagrams to relate to the electronic mail system. Performance accuracy as assessed on three experimental tasks determined the degree of an individual's success in training. A significant interaction between training approach and learning style was hypothesized

Table 15: Bostrom's Research Framework—Variables in the Framework

Trainee Characteristics	Training Environment	System Software
Mental Model of Systems Motivation to Learn/Use System Task importance Ease of Use Usefulness Enjoyment of Using Cognitive Traits Learning Style Visual Ability Procedurality Motivational Traits Self-Concept Need for Achievement Attitude Toward Computers . . . . Task Domain Knowledge Other Factors Previous Computer Experience	Methods: Conceptual Models Motivational Planning/ Management Physical Aspects	Ease of Use Type of Language (e.g., 4 GL) Type of Interface Direct or Indirect Manipulation

and supported. The researchers stated that the "efficiency of training models depends on a fit with learning style."

Fauley (1991), in an article in *Computing Canada*, argued that "capitalizing on learning styles will broaden the appeal and effectiveness of computer-based training (CBT)." Computer-based training should appeal more to Assimilators and Convergers than to Divergers and Accommodators. The latter two groups look more to personal experience or to other people for learning rather than relying on one's own analytic ability. Such learners would indeed seek human intervention and group discussion in order to learn effectively. Fauley also proposed that for Divergers and Accommodators, CBT may be "palatable" if human interventions were built into such a course.

Substantial research into learning style and the training of end-users (to use a new customized system or to use generic software tools) has been carried out by Bostrom and Olfman (1987, 1990). Bostrom developed an elaborate research framework identifying a number of variables for consideration in end-user training (see Table 15). It presents a conceptualization of what is involved when someone learns to use a software tool. Note that learning style is a (significant) component of that framework.

A central concept in Bostrom's framework is that of a "mental model," which a user forms about a system. Such a model is the source for his/her understanding of the system and of his/her development of facility with it. Learning is viewed as a process of model enhancement, such that each version reflects a more adequate understanding of the underlying software. A user can form a mental model via usage, analogy, or specific training. Thus "training can expedite the formation of mental models" (Bostrom & Olfman, 1990). Two main training approaches are instructor-oriented and exploration-oriented (e.g., CBT). Such approaches may make use of conceptual models, which may be either analogical or abstract.

Bostrom also identifies two types of training outcomes: *understanding* and *motivation to use the system*. Correct understanding reflects the correctness of the subject's mental model.

The above overview is outlined below to give the reader a "feel" for the level of complexity at which user training research has arrived. In their 1987 discussion, Sein, Bostrom, and Olfman advise:

- In EUC training, especially for novices, initially focusing on the development of appropriate motivational levels may be more important than knowledge acquisition.
- During training, the level of motivation is increased by providing relevant problems and ample feedback.
- When a manager begins to learn a software tool, it is important that there be a match between the level of motivation going in and the training methods used.
- Motivational planning must be applied to the design of learning tasks.

Certainly, learning style, as a key variable in the framework, would need consideration in order to achieve both efficiency and effectiveness in mastering of the software skills and in maintaining the motivation for the software's ongoing use.

Thus, in their 1990 work, Sein, Bostrom, and Olfman focus specifically on "The Importance of Learning Style in End-User Training." In addition to addressing learning style, the authors state that "**emotional or motivational states can have a tremendous impact** on the learning process and outcomes" [bold added], alluding to the need for "emotional intelligence" on the part of IS in the task of user training. They also declare that "research in instructional psychology has demonstrated that adapting instructional methods and teaching strategies to accommodate key individual differences including learning style has led to improved performance. They also refer

Table 16: Training Methods (from Bostrom & Olfman, 1990)

Learning Style	Conceptual Model	Training Methods
Converger	Abstract	Application-Based
Assimilator	Abstract	Construct-Based
Diverger	Analogical	Construct-Based
Accommodator	Analogical	Application-Based

to Kolb's implication that a subject may prefer one learning style in one situation and a different style in another. However, they assert that "even if learning style varies with situations, it will remain constant within a particular context," such as software training. They also caution that preference for a particular style is not synonymous with the ability in that style and a much lesser ability in others.

Sein, Bostrom, and Olfman, in the same paper, go on to refer to several research studies on the role of learning style in EUC software learning. These studies applied Bostrom's framework for EUC training. In each study, subjects were administered the KLSI prior to training. In general, abstract learners (Convergers and Assimilators) performed better than concrete learners (Divergers and Accommodators), and in most cases, active learners performed somewhat better than reflective learners. The authors conclude that "training methods need to be tailored to individual learning modes, but precisely how and under what circumstances is not entirely clear."

The authors do recommend that concrete learners be provided with analogical conceptual models while abstract learners would benefit more from abstract models. Reflective observers (Assimilators and Divergers) may require generic, pre-defined problems to solve together with instructions (construct-based training). Active experimenters (Convergers and Accommodators) could benefit more from working on their own (rather than generic) problems while being guided through their problem solving (applications-based training). Table 16, from Bostrom and Olfman (1990), summarizes the above recommendations based on training methods.

In the same article, the authors have included a fairly thorough list of *Prior Studies on Individual Differences Associated with Learning About Software*, for the interested reader.

It now remains a challenge for further research and development efforts to corroborate, adjust, or challenge the above proposals and to produce competently actual examples of training programs that incorporate the above advice.

Bohlen and Ferrat (1997) conducted an experiment involving 120 students in an introductory university computer course and applied either the lecture method or the computer-based training method for learning WordPerfect 5.1. Outcomes were measured by test scores, practicum scores, and assignment scores. All students had completed Kolb's LSI questionnaire and were then matched with a learning style.

Results indicated that Divergers, Convergers, and Accommodators did better using CBT, whereas Assimilators using the lecture method did approximately the same as those using CBT. Furthermore, Convergers were most satisfied with CBT and least satisfied with the lecture mode.

With expansion into intranets, extranets, and e-commerce support systems, and with increasing availability of and capacity for multimedia, learning of systems will take on additional importance. With multimedia, design of creative training programs will be encouraged, yet not all users will learn in the same way. Acknowledging this will be not only wise, but necessary.

It is indeed hoped that when the IS profession moves more consciously into psychological awareness, this will motivate widespread research efforts applying the psychology of learning in IS, not only to users but to all involved in the conception, development, implementation, and use of information systems.

It is further hoped that such new and impactful research will not be oriented mostly for journals and conferences, but that resulting development efforts will be undertaken, often jointly between industry and academe, to make all learning more effective, efficient, and satisfying.

FOR THE BEGINNER

At this point, you may likely be familiar with your Myers-Briggs type. From this you may wish to examine the derived learning styles of McCaulley and Natter, discussed earlier (IN, EN, IS, ES). You may also have become aware of your cognitive style. Combine the above information with recalling the last time you learned something substantial at work (e.g., new methodology, software tool, or the workings of an existing system). Can you see a preferred style in operation? Can you see any elements of other styles as well? Perhaps you can now consider earlier learning experiences, possibly in a post-secondary institution. What were your favorite subjects? Why? How did you learn them? Did this learning energize you? What were your least-favorite learning experiences? Can you relate learning style here?

Now, you may wish to consider the more elaborate system of Kolb. If you have the opportunity and the inclination, taking the formal Kolb Learning Style Inventory (KLSI) may be beneficial. This consciousness can then be related to specific IS learning instances. Are there more effective ways in which you could have learned what you needed to? Can you verbalize these alternative learning modes? You may, at this point, wish to discuss your insights with work colleagues.

Some motivated groups can take this awareness quite far, initiating more effective and synergistic learning. Also, another dimension to consider is the training of users. This includes design and development of appropriate training materials (which may include multimedia). What learning style(s) are your materials addressing mostly? For what styles might such materials be substantially lacking? Why? When testing the training materials on a sample of users, you may wish to address their learning styles, more formally or informally. Evident concerns can be documented. Of course, such explicit and substantial involvement of the psychological factor of learning style in the course of regular IS work may well depend on the maturity level of the IS organization regarding psychological awareness (see the proposed growth stages in Chapter VI). Eventually, it will be up to IS professionals, communicating as a community, to bring the issues of cognitive, creativity, and learning styles to a level of functional consciousness.

CONCLUSION

Personality type classifications attempt to describe how we function mentally and emotionally. They may also offer insights into why we function as we do. Cognitive and learning style identification focuses more specifically on the mental processing of information. It can be viewed as a "zoom lens" on one aspect of personality, which can be very significant in IT.

The material presented so far provides a somewhat broad and thorough orientation to the involvement of personality and cognition consciousness in IT. But need we stop here in considering relevant psychological factors? Is there yet another psychological dimension to the human being? Can this dimension have even a dramatic impact on the consciousness of the IS profession in this 21st century? The next chapter attempts to deal with this issue with creativity and challenge.

REFERENCES

Agor, W.H. (1991). The logic of intuition: How top executives make important decisions. In J. Henry (Ed.), *Creative management* (pp. 163-177).

Amabile, T.M. (1992). *Growing up creative: Nurturing a lifetime of creativity.* Creative Expression Foundation.

Amador, J.A. (1977). *Information formats and decision performance: An experimental investigation.* PhD Dissertation, University of Florida.

Bariff, M., & Lusk, E. (1977). Cognitive and personality tests for the design of management information systems. *Management Science, 23*(8), 820-829.

Benbasat, I., & Schroeder, R. (1977). An experimental investigation of some MIS design variables. *MIS Quarterly, 1,* 37-49.

Benbasat, I., & Taylor, R.N. (1978). The impact of cognitive styles on information system design. *MIS Quarterly, 2,* 43-54.

Bohlen, G., & Ferrat, T. (1997). End-user training: An experimental comparison of lecture versus computer-based training. *Journal of End-User Computing, 9*(3), 4-27.

Bostrom, R., Olfman, L., & Sein, M. (1990). The importance of learning style in end-user training. *MIS Quarterly, 14*(1), 101-119.

Chu, P.C., & Spires, E.E. (1991). Validating the computer anxiety rating scale: Effects of cognitive style and computer courses on computer anxiety. *Computers in Human Behavior,* (July), 7-21.

Churchman, C.W., & Schainblatt, A.H. (1965). The researcher and the manager: A dialectic of implementation. *Management Science, 11.*

Couger, J.D. (1996). *Creativity and innovation in information systems organizations.* Danvers, MA: Boyd & Fraser.

Couger, J.D., & Higgins, L. (1997). Comparison of KAI and ISP instruments for determining style of creativity of IS professionals. *Proceedings of the 28th Hawaii International Conference on System Sciences.*

Davis, D., & Elnicki, R. (1984). User cognitive types for decision support systems. *Omega—International Journal of Management Science, 12*(6), 601-614.

Ellis, D., Ford, N., & Wood, F. (1993). Hypertext and learning styles. *The Electronic Library, 11*(1), 13-18.

Epstein, M. (1996). *The role and worldview of systems designers: A multi-method study of information systems practitioners in the public sector.* College of Commerce, University of Saskatchewan.

Fauley, F. (1991). Learning styles have an impact on computer-based training. *Computing Canada, 17*(18), 34.

Ford, N. (2000). Cognitive styles and virtual environments. *Journal of the American Society for Information Science, 51*(6), 543-557.

Gardiner, M.M., & Christie, B. (Eds.). (1987). *Applying cognitive psychology to user-interface design.* New York: John Wiley & Sons.

Glass, R. (1995). *Software creativity.* Englewood Cliffs, NJ: Prentice-Hall.

Hayes, J., & Allinson, C.W. (1998). Cognitive style and the theory and practice of individual and collective learning in organizations. *Human Relations, 51,* 847-871.

Huber, G.P. (1983). Cognitive style as a basis for MIS and DSS designs: Much ado about nothing? *Management Science, 29*(5), 567-579.

Huitt, W. (1992). Problem solving and decision making: Consideration of individual differences using the Myers-Briggs Type Indicator. *Journal of Psychological Type, 24,* 33-44.

Huysmans, J. (1970). The effectiveness of the cognitive style constraint in implementing operations research proposals. *Management Science, 17*(1), 92-104.

Kirton, M.J. (Ed.). (1989). *Adaptors and innovators: Styles of creativity and problem solving.* London: Routledge.

Kolb, D.A. (1984). *Experiential learning.* Englewood Cliffs, NJ: Prentice-Hall.

Korthauer, R.D., & Koubek, R.J. (1994). An empirical investigation of knowledge, cognitive style and structure upon the performance of hypertext task. *International Journal of Human-Computer Interaction, 6*(4), 373-390.

Lindsay, P.R. (1985). Counseling to resolve a clash of cognitive styles. *Technovation, 3,* 57-67.

Lusk, E., & Kersnick, M. (1979). The effect of cognitive style and report format on task performance: The MIS design consequences. *Management Science, 29,* 787-798.

Marzano, R., Brandt, R., Hughes, C., Jones, B.F., Presseisen, B.Z., Rankin, S., & Suhor, C. (1988). *Dimensions in thinking: A framework for curriculum and instruction.* Alexandria, VA: Association of Supervision and Curriculum Development.

Mason, R.O., & Mitroff, I.I. (1973). A program for research on MIS. *Management Science, 19*(5), 475-487.

Matta, K.F., & Kern, G. (1991). Interactive videodisc instruction: The influence of personality on learning, *International Journal of Man-Machine Studies, 35*(4), 541-552.

McCaulley, M., & Natter, F. (1980). *Psychological (Myers-Briggs) type differences in education.* Center for Applications of Psychological Type.

Nagasundaram, M., & Bostrom, R. (1995). The structuring of creative processes using GSS: A framework for research. *Journal of Management Information Systems*, *11*(3), 89-116.

Palmquist, R., & Kim, K.-S. (2000). Cognitive style and on-line database search experience as predictors of Web search performance. *Journal of the American Society for Information Science*, *51*(6), 558-566.

Robey, D. (1983). Cognitive style and DSS design: A comment on Huber's paper. *Management Science*, *29*(5), 580-582.

Sein, M., & Robey, D. (1991). Learning style and efficacy of computer training methods. *Perceptual and Motor Skills*, *72*, 243-248.

Sein, M., Bostrom, R., & Olfman, L. (1987). Training end-users to compute: Cognitive, motivational and social issues. *INFOR*, *25*, 236-255.

Shneiderman, B. (1997). *Designing the user interface*. Addison-Wesley.

Spence, J.W., & Tsai, R.J. (1997). On human cognition and the design of information systems. *Information & Management*, *32*, 65-73.

Steinberg, R. (1997). The concept of intelligence and its role in lifelong learning and success. *American Psychologist*, *52*, 1030-1037.

Streufert, S., & Nogami, G.Y. (1989). Cognitive style and complexity: Implications for I/O psychology. *International review of industrial and organizational psychology*. Chichester: John Wiley & Sons.

Tan, B., & Lo, T. (1991). The impact of interface customization on the effect of cognitive style on information system success. *Behaviour & Information Technology*, *10*(4), 297-310.

Williams, R., & Stockmyer, J. (1987). *Unleashing the right side of the brain—The LARC creativity program*. Stephen Green Press.

Witkin, H.A. et al. (1971). *A manual for the embedded figures tests*. Consulting Psychologists Press.

Chapter IV

The Deepest Inner Self:
A Foundation for "Emotional Intelligence"

INTRODUCTION

Until now, we have examined how we behave and function, and also how we think and learn. But, are we simply "rational animals" with an intellect, feelings, and a body? Is there, from a conceptual as well as experiential point of view, yet another component to the human person?

This chapter assumes that, indeed, the deepest inner self (inner core, center, being, soul), does exist in and can be consciously accessible to a well-adjusted, aware human person. Furthermore, this central core can truly provide rejuvenated psychological energy in times of stress and change, and it can also provide stability and impetus to significantly creative efforts. Thus, it is proposed here that conscious awareness of one's deepest self can indeed add a very important dimension to the work of an IS professional, particularly one whose work involves human interaction. As well, it is pointed out that connection to one's inner self provides the basis for the currently popular notion of "emotional intelligence" (which will be defined later in this chapter), both at work and in personal life.

As such, this chapter is itself an example of an attempt at generative creativity, whereas previous chapters have perhaps been more adaptive and less "daring." Here, we examine a perspective on the key components of a human person and on the relationship between them. We touch upon psychology (in the philosophic viewpoint), human spirituality, motivation, and the dynamics of human relationships. We consider limiting and addictive behaviors and dysfunctional attitudes, individually and organizationally. In this chapter, the quotation from Poincaré, *"Creativity is the fruitful combining which reveals unsuspected kinship between facts long known but wrongly believed to be strangers to one another,"* is likely most applicable. Above all, this chapter promotes learning to live and work from one's inner core, rather than exclusively from the intellect (and perhaps some emotion). Such an orientation can provide the IS professional of the 21st century with inner freedom and emotional empowerment that can surpass by a "quantum leap" the benefits from other levels of psychological awareness. Such an inner shift in one's essential consciousness may be not only desirable, but essential for the long-term survival and fulfillment of the modern IT worker.

While openness to what many may feel "new" or "radical" in this chapter is encouraged, so is a healthy skepticism. However, based on theoretical and anecdotal writings as well as personal experience, I am convinced that addressing the inner center in a book on psychological growth for IT workers cannot be ignored.

A FOCUS ON THE DEEP INNER SELF

In the '60s, French educator André Rochais noticed that some school children were optimistic, resilient, curious, and energized, while others, of equal intellectual capacity, displayed few such characteristics. He then set out on a personal mission to determine what, within a child, has to be specifically addressed in its upbringing so that it might develop into a person who is "significantly alive." Rochais observed, intuited, and read the works of numerous psychologists and educators. In this search, he was particularly influenced by the work of psychologist Carl Rogers. In his classic, *On Becoming a Person* (1961), Rogers points out the existence of a basic positive "core" within every human being. His main theme seems to be that at the root, every person is basically good. However, many can go through life essentially unaware of this basic goodness and dynamism for life. This basic goodness is primarily evoked to consciousness by vivifying relationships.

Rogers points out: where the "hearts" of persons are not engaged in a relationship, there cannot be much true, deep growth. An argument he puts forward to support his claim is that psychotherapy cases in which the therapist was distanced from the client and dealt with him "clinically" have not shown intended results. Rogers pointed out that this involvement of the "deeper self" is vital to all fruitful human relationships, not only to effective psychotherapy. In relation to work communication, Rogers noted:

"I have found that the more genuine I can be in the relationship, the more helpful it will be…It is only by providing the genuine reality which is in me, that the other person can successfully seek for the reality in him…It seems extremely important to be real."

Through considerable observation, synthesis of modern hypotheses with ancient wisdom, and increased self-awareness, André Rochais came to isolate and name a specific component of the human person in addition to intellect, feelings, body (with its senses), and will. He called this part the *inner being*, the deepest energy of pure essence, of "felt truth," where traits such as patience, motivation, confidence, harmony, clarity, and desire for truth are found. He noted, as an educator, that it is this inner being that must be addressed adequately for a developing person to become resilient, radiant, and creative.

We can often recognize this inner being or deepest self as a "felt truth" or "gut level knowledge." When asking, "Of what about myself am I absolutely sure?"—"What trait of mine do I 'own' with absolute certitude?"—we may come up with a deeply felt (often actually felt in the abdominal area) conviction, such as, "I am a good teacher/analyst/writer" or "I am a very sympathetic person." We can make such statements with such poise and certitude (while not being boastful) that we recognize this "knowledge" is not emanating merely from the intellect, but from a deeper and more "grounded" part within us. Here is our deepest self at work. Other people may recognize the being as a conviction so deep and so true that it gave them strength and stability in a life crisis situation. Yet, not everyone has had a conscious experience of his/her deepest self. For many, the intellect/feelings/body form the sole reality of their life experiences.

According to Rochais' framework, this inner being is intended to be the center, the dynamo of a person's psychological functioning. It (the being) is restful, consolidating, vivifying, and fulfilling. The other human components, i.e., intellect, feelings, body, and senses are to be connected to the being, and to receive life energy from it. In such a life experience, the intellect receives from, reasons, and verbalizes the felt truth of the being; the feelings vibrate the being; and the body encases the being and is re-energized by it. Such is the experience of "personal wholeness."

However, for most people, this deep true self (inner being) is only dimly in their awareness, if at all, on a daily basis. Yet, it is at this level that one can feel the most powerful energy source for handling change, stress, and uncertainty with perseverance, calmness, and even enthusiasm. To those who have experienced it and can connect to it inwardly at least somewhat regularly, the deep self is a real, distinct energy, which if allowed, will integrate all other human capacities and allow them to operate in harmony. As mentioned earlier, Couger and Zawacki have noted that system development professionals showed a lower-than-average need for socializing, but a significant need for growth. In their time and context, "growth" may have largely referred to professional competence. Today, however, interdisciplinary considerations are growing and becoming accepted. Just as an Olympic athlete may consider nutrition, biochemistry, kinesiology, psychology, and philosophy to deliver an optimum performance, so might the "modem" systems professional examine the possibility of discovering a deeper psychological reality within, as part of his/her professional and personal growth. However, for the concept and the reality of the deepest self to be "palatable" to many IS professionals, it will need to be formulated and presented perhaps not so much "artistically" as in a mental framework that is familiar to IT.

The Two-Tiered Model of the Human Person

For a majority of people, at least in today's Western world, their self-awareness is largely limited to a one-tiered view as shown in Figure 3. Here the mind (intellect), feelings (emotions), and the body (with its senses) are identified. The will, then, is the force that channels one's consciousness to the various parts in various proportions during the day.

A person at this level of awareness has the three components available for use. However, each of these three parts can be at different levels of "positivity" or "negativity" at any point in time. For example, the mind can be thinking clearly or be confused; feelings can be elated, peaceful, or angry; the body can be healthy or sick. Indeed, these very "human," imperfect components

Figure 3: One-Tier Model of the Human Person

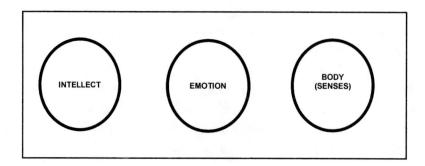

Figure 4: Two-Tier Model of the Human Person

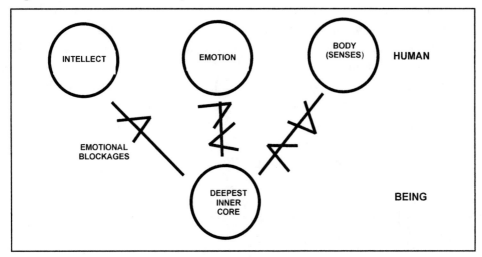

cannot be counted on to come through for us "in top form" in situations of definite need (e.g., dealing with a system crash the day before installation).

However, a variety of personal growth programs and literature, some of which have received notable acceptance from professional circles, propose the *two-tiered model* of the human person, shown in Figure 4. Here the three human dimensions are receiving energy/direction from the deepest self/core energy/inner being, which operates on a separate, deeper level. Moreover, the substance of this core energy is *only positive*—there is nothing negative and nothing missing (for deep psychological wellness) in the inner being. The difficulty, though, lies in establishing and maintaining a strong enough connection between the *human* and the *being* parts.

Dr. Bernie Siegel, well-known health counselor and former Yale University surgeon, points out in *Peace, Love and Healing* (1989):

"Most of us are in touch with the inner self only intermittently, if at all…it's not located in the conscious mind…this perfect core self seems to fit in with neuropeptide theories…I just got to know this perfect core self through meditation…however you get there, you'll know you've arrived—it's like coming home, and home is where the healing can begin…within your true, unique and authentic self."

It is worth noting that the role of the will (the "channel changer") is different in the two tiers. We can usually switch our attention, our concentration, willingly from the mind to the body or to the feelings—the use of will is more direct here. However, *it is not possible to will oneself into the deepest self.* One can, at best, willfully relinquish control of the mind, body, or feelings and try to focus on one's "felt truths"; the core energy simply surfaces to consciousness.

Also, in the two-tiered model, note the "wedges" between the human parts and the inner core energy. These are the reasons why most of us are not often in touch with our "true, authentic self." According to a number of personal growth perspectives, there are *five basic inner needs* that we have when we are developing psychologically:

1. To be SEEN for what we are experiencing deep within us.
2. To be HEARD attentively when we try to communicate something of deep importance to us.
3. To be TRUSTED in our innate ability to develop "for the good" without "micro-managing."
4. To be RECOGNIZED and addressed as a unique person who has infinite value much beyond behavior and performance.
5. To be SAFE to open deeply inner experiences, awareness, struggles, and discoveries without being judged.

While likely no one has had these needs perfectly met, an adequate satisfaction of the above five needs acts on the psyche like water on a growing seed. This nurturing is what allows the core energy of the inner being to begin to surface to the consciousness of the intellect, feelings, and body. A person growing in such an environment is on his/her way to becoming integrated and "whole."

Significant frustration of the above five needs, by deprivation and/or negative (perhaps, but not necessarily abusive) impact, creates emotional blockages, many of which are initially subconscious. These are the wedges of Figure 4. These are largely the reasons why many people, IS professionals included, are not sufficiently aware of or connected to their most positive inner asset. Personal growth programs aim at recognizing, identifying, and removing these debilitating and limiting blockages.

The two-tier model admits to the existence and the central nature of the inner core in an integrated, functional *human being*. It is much like application software, on one level, running under an operating system, on another level (an analogy amenable to IT). It is like having lived in a one-story house and now realizing there is an elaborately furnished basement.

We have seen that the deepest self/inner being/core energy is very em-powering and vivifying, yet one cannot will himself into it. We have seen that it is separate from the intellect, feelings, and body, but gives essential "tone" to all three. At this point in the presentation, skeptical, analytic IT minds may be looking for more facts, theories, and empirical evidence regarding this proposed level of psychological functioning. The following sections may be of help.

PRH PERSONAL GROWTH PROGRAM

After identifying the inner being as the component of the person that needs to be addressed during upbringing for "fullness of life" in adulthood, André Rochais founded PRH (Personnalité et Relations Humaines) in France in the early '70s. PRH is an international school of psycho-pedagogy, now found in more than 32 countries, where members are dedicated to fostering personal growth. PRH provides a methodical formation approach to help people know themselves better and to become more fully who they really can be; to overcome problems and inner emotional blockages that impede their growth; to free their life forces and creativity; to improve their relations with others; and to take root in the depths of themselves, thus living in touch with an Absolute.

PRH provides workshops (many taking 30 hours), follow-up in small groups, personal guidance by an accompanist, and printed "observation notes" to correspond to various workshops. It is emphasized, however, that PRH growth happens mostly through inner experiences rather than simply through analysis and adoption of a different mindset. Thus, "PRH growth and development is first and foremost an experience to be lived, rather than a teaching to receive." Apart from observation notes (consisting of only a few pages per workshop), PRH-International has produced only one book, *Persons and Their Growth* (1997), which summarizes key elements of the PRH philosophy.

For the IS professional, PRH concepts can be viewed as *a systems analysis on the inner functioning of the human person*. (It may be quite likely that much of their writings were compiled by an NTJ personality type.) The writings are structured, often in point form; they are deep and wholistic, involving the mind, feelings, body, and above all, inner being; they are practical, and based on considerable empirical evidence to back up the presented viewpoints. In the writings, PRH deals with the inner being in a very systemic and coherent way, rather than using a poetic, "touchy/feely" perspective. This is likely to appeal to the structured and thinking orientation of many IT workers.

PRH does have a website at *http://www.prh-international.com*, but it does not advertise and, as such, it is not well known. It seems to have genuine human development as its central mission, rather than expansion and proliferation. It does not operate to produce more effective businessmen/professionals, but to help businessmen, professionals, and others, within their own deep experiences, to become deeply rooted, balanced persons, who will then, without specific effort, become more effective at their work.

The following are noteworthy excerpts from various PRH Observation Notes, which are largely meant to supplement inner experience and self-discoveries {author's clarifying remarks are provided in brackets}.

On the goal of life:

"The goal of life is to be myself, that is to actualize the potentialities of my being, and at the same time to accomplish that for which I am meant."

"I must give full priority to knowledge of myself ... especially the depths of myself."

"... with most men and women, this innermost self is buried in the unconscious ... To become aware of it ... entails a long process of hard work ... with most people, their inner being must be set free."

On the source of one's security:

"We can build our personality on peripheral gifts and relate to others {co-workers, end-users} on that basis: intellect (by pursuing studies), action (by being enterprising), aptitudes in a profession (becoming competent).

... and so our personal development is based on our peripheral gifts ... but the essential deficiency {unawareness of the deepest self} remains {inner emotional blocks from not being seen, heard, trusted, recognized, or safe enough—blocks that made our inner being inaccessible.}

... when we fail socially {or in academics, profession or relationship} we fall apart" {because our inner being has never emerged to our accessibility}.

"It is of primary importance to turn to the past to unearth and identify the original deficiency" {blockage from the being because of lack of attention to its development}.

"The need for attachment {to job, status, spouse, etc.} is not the most essential need" ... What is fundamental to everything is the need to exist and be oneself {the self of the being}. From this stems the need {especially in development years} to be helped and encouraged to exist according to who one is {at the level of deep self-being and not solely at peripheral levels}.

On the deep inner self:

"Deep within us at the very heart of our being we can discover the positive aspects of our personality. It is these positive aspects that make us worthwhile persons."

"Zen, Yoga, Vittoz, etc., prescribe exercises to help decrease cerebral functioning. Other exercises increase the awareness of sensations. All of these facilitate access to the reality of the positive self. For people not accustomed to internalizing and for people who function on the predominately cerebral level {e.g., NT and many SJ types in IT}, these exercises are almost indispensable."

"Little by little, the eyes of the intellect will become accustomed to this unknown country ... and one day it will realize that it is its homeland ... the intellect will then adhere to it, seeking henceforth at this level of being the truth on all things: the intellect will become docile to the inner being, whereas before it, it unwittingly tried to direct it."

"This positive being is a source of life ... we are caught up in a life-force which is calm and powerful at the same time, and which can lead us to a certain limitlessness of self; this experience of the infiniteness of our being is very dilating and frees us from the narrow boundaries of everyday living ...; this is the starting point of true liberty ... ."

"For a long time, the scientific approach was embarrassed by the observation that love is the central force in every person. We were afraid of giving in to a simplistic or sentimental description of the human being ... Our observation and research oblige us, nontheless, to surrender to the fact

that such is the case. The center of the human being is love; love is the most powerful reality in us."

{Then, PRH provides a much needed clarification:}

"There are, however, two levels of love: i) love that is emotional, instinctive, impulsive, passionate, sentimental—this love is better known; it is love in its sensitive functioning. ii) deep love ... it is calm, vast, powerful—this is love in its in-depth functioning."

"To become conscious of the deep love in our innermost being is of capital importance for our growth ... living it fully leads to complete self-actualization."

On working and deep rootedness:

"Some use their gifts but do not live them ... these are people of action {or intellectual analysis} who do not take time to stop and commune with their source of being and life; cut off from their inner source, they ... wear out; they live dissatisfied, ever seeking on the outside what they unknowingly carry within." {Is this becoming increasingly common in IT?}

"Those who are attentive to the innermost zone discover it is a fertile field of creativity and intuition."

> *Quotations were taken from the following PRH Observation Notes: "The Goal of Life" (1983), "Deficiencies Might Be Relieved Without Being Removed" (1983), "My Affective Life" (Second Commentary, 1983), "The Need to Attach Oneself" (1985), "The Positive and Innermost Zone" (1983), "Aiming at the Humanization of the Business Firm" (1985).*

From the above quotations (and other PRH Observation Notes available from the organization), the reader may notice a focus on a deep existential human reality expressed in a thoroughly analyzed, structured, and coherent manner.

Several points may be noteworthy for the IS professional. Firstly, there may be individuals who, in their growing years, had sought inner security in the structure and sense of control that computing had provided for them. They, thus, built their personalities and their lives largely around their peripheral talents for mathematics, computing, and analysis. Yet, unknown to them, they had experienced a deprivation of attention to their true, deep self, which thus had not emerged significantly to their consciousness. Current work realities have brought them constant change, uncertainty, and the unpredictability of human relations. Without inner access to their deepest self, which remains solid in the midst of change, such people are rapidly burning out, without awareness of an indwelling remedy. (In a short Internet poll of 250 IS professionals in the year 2000, 48% responded being "close to burnout," answering at least five on a 7-point scale.)

Another point relates to a specific event. At the 1995 National Convention of the Canadian Information Processing Society, a keynote speaker (a management consultant) made some "breakthrough remarks" addressed at IT workers. He noted that, in the past, considerable communication with others, especially users and higher management, was carried out from a perspective of power and confrontation. Instead he proposed that IT professionals learn to communicate from a perspective of love (he actually used that word, at a "techie" convention). For this promotion of synergy, the speaker received a standing ovation. Something was realized and, at least on some level, accepted.

However, when working from a "one-tier model" of the human person, it is indeed hard to understand *how* to communicate "with love." With only the intellect, feelings, and body to work with, the predominately thinking and structured IT work world will interpret a call to love as a call to shift from the security and comfort of one's intellect to the undefined and questionable world of feelings. As a result, many IS workers would resist and even ridicule a prescription to "love one's users" which they would view as a prescription to get "touchy and feely."

However, with recognition of the call to love as a call to live in connection with one's deepest inner core energy, i.e., a call to in-depth rather than sensory love, as clarified in PRH, the IT worker, upon understanding this, would no doubt be less threatened and perhaps even more encouraged to "take another look" (at least to the point of considering following the PRH excerpts in this section).

Another noteworthy and indeed relevant point is that some people "use their talents without living them." *Living* one's IT talent would refer to programming, designing, or IS planning, while simultaneously being (consciously) connected to one's deepest inner self (core energy). In this state there is much more inner freedom and resilience to rebound from crises or disappointments at the level of one's work.

Finally it is worth noting that the deepest inner self can indeed make possible significant intuitive insights and creative efforts. Furthermore, the energy of the being within can provide one with the inner strength to persevere so as to make one's intuitive hunches reality.

At this point, two key questions may arise in the reader's mind. First, is the energy of the inner being indeed a tangible (although not easily accessible) human reality which operates on another deeper level than the familiar intellect, feelings, and body? Secondly, is the IS profession, particularly in the Western world, ready for addressing and promoting the awareness of such a reality?

Two more excerpts from PRH Observation Notes can address, at least, the second question. The first deals with the readiness of the Western world to "go deeper within," while the second addresses the relevance of PRH growth and awareness in business firms and thus in IS departments.

"Today, we are standing on a...threshold; this is developing in societies which[:]

- *have entered a state of material prosperity,*
- *have greatly developed the intellect,*
- *are starting to have a taste for the interior,*
- *possess a developed language permitting the expression of inner realities,*
- *are experiencing a great development in the psychological sciences,*
- *are finding that progress in the sciences and technology creates new problems in the face of which human beings feel helpless {e.g., stress in IT work},*
- *are finding that ever broader fields of knowledge are not sufficient to satisfy the deep aspirations of the person, and*
- *are starting to discover the extraordinary wealth of potential, and therefore of creativity, that lies dormant in the depths of each person."*

"We who have carried the PRH experience very far can testify that incredible riches have been discovered within ourselves and have been actualized. We found much more within us than we anticipated...and yet, we are but ordinary men and women. It is on the basis of this...that we dare to say: our Humanity is extremely underdeveloped in the area of in-depth riches."

PRH has not ignored life in the business environment. In its 1986 observation note, "Aiming at the Humanization of the Business Firm," we note the following:

"If the business firm opts for the growth of its personnel, it ought to favor the integral growth of the person, not only the professional growth which gives immediate results.

Integral human growth...is a way of liberating all the riches of being, frequently buried in the unconscious and/or paralyzed by past hurts.

At the level of the Being {core energy of the deep inner self}, there is everything that is needed for persons to create harmonious relations with those around them and to cooperate effectively in a common task. There is everything needed so as not to capsize under the pressures {including work stresses} of the hard blows of life, and to restore a solid footing within oneself so as to carry on with a firm step."

"The chief executives of a firm {including CIOs} and all those who have influence on the orientation of its policies can broaden their horizons...by fostering integral human growth and not only professional growth; firstly, though, the chief executive and persons in {IT} management must choose this integral growth for themselves and experience its benefits."

Noteworthy indeed is the comment that at the level of the deepest inner self, we all have the necessary inner resources to maintain harmonious relations with co-workers, to cooperate effectively in work endeavors, and to maintain poise and stability in times of stress, without "capsizing." And, today, among the chief "soft topics" of interest to IT executives, we see concerns about user relations, effective teamwork, and prevention of burnout.

As we shall soon see, a number of other authors/teachers are also referring, in their own vocabulary, to the deepest core level of the person and the practical psychological benefits of "being in touch." Thus, at least a (growing) segment of Western society has been ready to transcend the intellect (but of course not to deny or diminish its value) and move towards a rooted, personal integration.

In light of this, perhaps the above excerpts from an established and successful worldwide personal transformation effort will be a first step in legitimizing the considerations of this dimension of human psychological growth for greater effectiveness among IT workers. Anticipated initial skepticism, apprehension, and even cynicism regarding the inner self is not likely to be so much because of the idea or possible inner experience, but because of preconceived notions that "soft" skills are not "strong," are not "scientific," and are "too wishy-washy." Perhaps the volume of PRH materials (including those referenced but not quoted), with their thorough, clear, practical, and well-structured approach to life at our deepest inner center, will intrigue a fair number of thinkers as well as feelers within IT, motivating many to at least keep an open mind and to examine more evidence, some of which is presented below.

THE HOFFMAN INSTITUTE

We have seen that PRH developed in the early '70s in France and promotes the view and the experience that in addition to a body, feelings, and intellect, we possess a genuine, deep inner energy that is fulfilling and totally positive. This is our true essence with which we are called to connect.

In the late '60s, western U.S. psychologist Bob Hoffman developed a "quadrinity process" to help guide clients to freedom from compulsive and self-defeating behaviors. He used the term "quadrinity" to refer to four interactive aspects of oneself: physical body, emotional self, intellectual self, and spiritual or higher self.

Hoffman's process, which deals mainly with identifying, expressing, and experientially evacuating blocked emotional energy resulting from inadequate parenting, initially met with skepticism and criticism. In time, however, it has repeatedly proven its lasting effectiveness and has been promoted by people in various walks of life. Today, the Hoffman Institute (*http://www.hoffmaninstitute.org*) is present around the world, offering guided workshops to liberate people internally. The Institute has a Web presence and has been praised highly by prominent scientists, actors, businessmen, and others, all former clients.

Hoffman pointed out that the four aspects of self are "interactive and form a complex feedback system." His Process aims to create a harmonious integration of the four aspects of self, and results in "total self-acceptance, self-forgiveness, and self-love."

It is very easy to see the parallel between PRH and Hoffman. Both were developed independently, around the same time. Both point to the deepest (highest) self as the greatest source of inner strength and stability. Following are noteworthy quotations from Hoffman's best-known book, *No One Is To Blame.*

"Everyone possesses positivity, no matter how negatively he or she behaves."

"The most negative trait possible is the inability to love oneself and others. If our parents did not know how to love themselves and others, they could not teach us how to give and receive love."

"Love is...a state of being, and it comes from the spiritual essence, 'the light' within ourselves."

"If we experience more than enough loving during those formative years, there will be a lifelong surplus we can share with others."

"Lack of self-love is like a locked cage, and intellect is not the key."

"The intellect is a mere speck in the ocean of emotion."

"...there are no tricks to the truth of self-love; meditation, psychoanalysis, jogging, drugs, encounter, fasting, massage, you name it, none of them will make a difference...until you learn to love yourself" {learn to connect to the deep, true inner self}.

"Negative love programming is usually below the level of awareness, and people cannot be ordered, coaxed, or persuaded out of it."

"Negative love programming is recorded in our emotions during childhood; it also affects our intellect and obscures our perfect spiritual self {core energy/inner being of PRH}."

"Until the adult intellect learns that the real source of its problems is the negative emotional child, it is powerless to change... ."

"The conflict we experience stems from the lack of integration between the present-day intellect and childhood emotions; reprogramming is to re-educate the emotional child within to drop its negative programming, grow up,

and join its spiritual self and intellectual self in present-day harmony" {a harmony without which the modern IT worker can hardly survive}.

"Head trippers at first sometimes have difficulty seeing with their inner eye."

"There's knowing and there's KNOWING...the real KNOWING doesn't need...explanations."

"The importance of discipline and self-control in achieving our positive goals is vastly exaggerated. If that alone could do it, we would all be living positive, loving lives."

"Change is possible...you have but to uncover your indestructible, diamond-like, positive, loving spiritual essence—{we can recall Bernie Siegel's comment here} the Light that is you."

Hoffman's Process is directed primarily at blockages from parenting that did not allow the deep self to emerge adequately; however, it is not limited or primarily aimed at those with severe traumas in early years. It can be viewed as a growth process rather than a therapy (and, according to Couger and Zawacki, IS professionals do thrive on growth).

The Quadrinity Process is usually given at designated centers in an intensive, week-long format with considerable "homework" before and after the week. Work can become intense and arouse powerful emotions, which are then evacuated for lasting change and increased, empowered authenticity.

Because of its impact, reputation (and possibly even its cost), the Hoffman Quadrinity Process has been dubbed "the Rolls Royce of personal growth programs." While Hoffman is challenging, the Process is very empowering, in the true sense of the word. IT workers often notice emotional drainage—the antidote to this, of course, is access to a deeper, authentic inner power that can refresh the spirit, re-energize the body, and refocus the mind. The Quadrinity Process makes such a power accessible.

Both PRH and Hoffman's Quadrinity share another common feature— their reference to the deep inner self (being, spirit) is coherent, structured, and well thought-out, in addition to reflecting over 30 years of empirical evidence. This is very significant for a large fraction of IT workers, who are "structured thinkers" by personality and who would likely not even consider such a growth path were it promoted to them by a "poetic" or "artsy" approach that is often natural to intuitive feelers.

Yet, in our current North American society, at least, the promotion of accessing and developing the reality of the deep self is indeed proliferating. An increasing number of teachers/authors are finding new words to speak of this inner dynamism to the "modern techno-centered man" who is often uncomfortable with the mystic, the poet, or the priest.

THE DEEPEST SELF - FROM OTHER CONTEMPORARY AUTHORS

A number of contemporary promoters of personal growth have expressed, each in his/her own way, their awareness of the deep, inner core energy that is the source of inner stability, resilience, and creativity.

Such persons, who have often synthesized inner experience with high-level, interdisciplinary academic development, have been expanding the norms of acceptability in Western culture on what it means to be human and to be truly educated. Speaking with clarity, conviction, and often intellectual rigor, they have been providing, through their generative creativity, a solid basis for many educated professionals to at least consider exploring this transcendent dimension of humanity with a healthy curiosity. Some such professionals may even make a commitment to personal and professional growth.

Following are relevant excerpts that emphasize the role and power of the deepest inner self:

Joan Borysenko, cellular biologist, psychoimmunologist, and former director of the Mind/Body Clinic at New England Deaconess Hospital, remarks:

"The experience of the Self requires making the mind one's servant rather than letting it be our master...when thinking is not required and the mind is allowed to rest, the Self is automatically experienced...when we experience ourselves as this essential Center rather than as any of our roles, we can function optimally, unimpeded by fears and desires...when we experience the Self...there is no fear, only love, peace, wisdom, interconnectedness, and an overwhelming sense of love and safety that abide in the now." (Guilt Is the Teacher—Love Is the Lesson, 1990)

Borysenko, a Hoffman Quadrinity Process graduate, points out how when we are connected to our Center:

"...colors seem unusually vivid, scents incredibly fragrant, sounds unusually rich, and textures almost alive...We can then perform our actions in the world with greater awareness."

The focus on greater awareness and on functioning optimally can indeed be of immediate relevance to the often overburdened IS professional. Yet, many are found to wonder *how, in an intellectually focused area such as IT, we can make our intellect "a servant, rather than a master"* so as to function optimally. Insight on this point is provided further in the chapter.

Personal development teacher and former university professor *Wayne Dyer*, in the introduction to a series of his taped presentations, remarks:

"I want to talk to the real you...not to the business or professional person you are, not even to the hobbyist...though they are all part of you...let me speak to that invisible, weightless part of you...call it your spirit or essence, if you like, or your psyche or inner being. You could even call it your soul...you know the part I mean...."

He goes on to declare in his series, *The Awakened Life:*

"Once you get tuned into the awakened life, success and performance and achievement matter less and less...yet they show up more and more."

In his book, *Your Sacred Self* (1995), Dyer asserts:

"Your interaction with the material world will be altered dramatically when you become self-aware."

Perhaps many IS professionals indeed wish that they could interact with the demands of their job with more composure, confidence. and optimism, even in the face of distressing realities. Some may indeed be motivated to explore the explicit connection between deeper self-awareness and job effectiveness seriously and systematically.

Dyer also states that while connected to the awakened life, one will live authentically and not have difficulty being oneself (that is, the deep, energized, positive, "true," self). It can indeed be an interesting and useful research question in IT management to explore what events and experiences might tend to block IT workers from accessing their deepest selves. What events in IS work give rise to specific emotions? Are there classes of IT workers that will tend to react with similar emotions in given work difficulties? Are such reactions related to personality type? What (subconscious) belief systems might initiate these reactions? Are such reactions "contagious" and what psychological conditions make certain IT workers more resistant to epidemics of negativity, blame, and resulting dysfunction? Perhaps a new field of IT management, called *Psycho-informatics,* may emerge (now, *that* would be generative creativity).

Popular U.S. psychologist *Phil McGraw* has recently been making a significant impact in the area of personal development. In his best-selling work, *Self Matters* (2001), McGraw promotes the view that one's life has a "root core," and once it is understood and accessed, this *unlocks a powerful force* within us—he then gives a nuts-and-bolts approach to discovering the "real you."

In each of the above-mentioned cases, as in Hoffman and PRH, it is not difficult to recognize a common thread. The recurrent idea of the pure, deep inner core energy is shown to have more potential for psychological empowerment in personal and professional life. Can IS professionals afford to ignore such a potential? Furthermore, if the majority of them has indeed been slow to take note, it is perhaps primarily because the mindset created by IT work (in the context of Western culture) is not (yet) able to assimilate easily the ways and language that has largely been used to promote the deepest inner self? Might the "two-tier model" and its analogy to operating system software below the application software be a step in bridging two paradigms?

IT workers, more of whom are TJ types in MBTI, are accustomed to structure and precisely, logically definable realities. For some, perhaps many, such realities have exclusively formed their life view. In such a view, being in control, understanding, and "scientifically predicting" are counted on to provide inner security, self-worth, and motivation. However, the rapid change in technology and in IT paradigms (such as OO), coupled with increased expectation in the Internet era of mega-connectivity, are eroding, at times significantly, "security from control." Consequently, many IT workers may, knowingly or not, be in or heading for a major crisis, not only due to overwork and frustration, but most importantly due to significant experiential threats to their source of inner security and identity on which most (perhaps all) of their life rests.

The IS profession seems to be recognizing the threat of impending "mass burnout." In a recent issue of the *Cutter IT Journal,* dedicated to IT burnout, Edward Yourdon (2002) states:

"Burnout is still a topic that most senior managers would rather not confront, but it has become so prevalent and severe that some IT organizations have become almost completely dysfunctional."

However break*down* can lead to break*through.* To minimize the damage of prolonged stress in IT, a new vision is first of all needed, a *wholistic model of a "functional IT worker."* From this vision, effective assistance measures could indeed be provided.

A major contribution to such a "new vision" that can apply to IS professionals can be seen in the work of *Gary Zukav,* author of the bestseller *The Seat of the Soul* (1990) and *Linda Francis*. In their collaborative book, *The Heart of the Soul* (2001), Zukav and Francis declare:

*"Spiritual growth is now replacing survival as the central objective of the human experience…The old species explored the physical world and it created security by manipulating and controlling what it discovered. The new species creates security by looking inward to find the causes of insecurity and healing them…This is the path to authentic power…**Authentic power is the alignment of your personality with your soul"** {i.e., deepest core energy} [bold added].*

Can not the above be the basis for a strategic vision of the "functional IS professional" of the 21st century? A vision statement can indeed be brief, yet profound and transformational. The symbols "$E = mc^2$" have made an enormous impact on human understanding and experience in the previous century. Can not the eloquently expressed realizations of Zukav and Francis, effectively operationalized, indeed make a significant impact on the potential for inner resilience, psychological robustness, and energized creativity for IT workers today and tomorrow?

However, strategies must lead to tactics. Vision must be implemented and tested. If the above statements can form a new vision for an "operational paradigm" for the IS profession, established and tested growth programs such as PRH and Hoffman's Quadrinity can guide tactical and operational realization of such vision. The tools are here! The next steps are *understanding, acceptance,* and *adaptation.*

Can the current IS professional expand his/her professional paradigm? Can he replace his source of security and motivation, which has largely been the capacity to understand structure, and control with a deeper source of security found on another "tier" within? Can the IS worker adjust to another layer below the intellect, a layer with vivifying core energy? Furthermore, can the IS professional develop a healthy curiosity and perhaps a new slant on his/her understanding of spiritual growth?

AN UNDERSTANDING OF SPIRITUALITY

Many hours can be spent (fruitfully or otherwise) in trying to define "spirituality." Yet, there is a number of emerging attempts to focus on "spirituality in the workplace." Journals are being established to this end, and conferences are being held. For example, at the Hautes Etudes Commerciales at the University of Montreal, conferences were held in 1998 and in 2001 on "The Integration of Ethics and Spirituality at Work: Examples of Different Traditions Throughout the World." These events brought together well-known international leaders. Topics presented and discussed have included, "Developing an appropriate spirituality for leading an organization" and "Spirituality: The next major challenge in management."

Yet, on the other hand, there is observable reservation, apprehension. or even cynicism as to attempting to relate spirituality to work. Many may fear indoctrination in a foreign and, to them, unacceptable life-view; some fear that attempts to spiritualize the work effort would bring even more conflict and tension. Indeed, in the current climate of psychological awareness in the Western world, introducing and promoting "spirituality" for wide acceptance and noteworthy change in effectiveness is a challenging task. Such a task requires patience, sensitivity, and above all, a thorough, scholarly understanding of the essence of human spirituality as opposed to emotional and cultural bias.

A unique, thorough, and soundly academic perspective on the nature of human spirituality is evident in the work of Daniel Helminiak. While the standard, readily acceptable model of the human person consists of body and mind, Helminiak, with doctorates in both theology and psychology, presents a tripartite model, identifying organism, psyche, and spirit (1998).

The *organism* refers to the physical life-form, the object of study in physics, chemistry, biology, and medicine—it requires satisfaction of life-sustaining physiological needs.

The *psyche* is presented as a dimension (but not the exclusive one) of the human mind, constituted by reasoning, emotions (feelings, affect), imagery, and memory. "Together, these determine habitual response and behavior, personality." The requirement of the psyche is to be comfortable, to feel good.

The *spirit* is presented as a "distinctively human dimension of mind, determined by self-awareness and experienced as spontaneous question, marvel, wonder, a dynamism open to all there is to be known and loved... the unbounded unfolding of spirit requires openness, questioning, honesty, and love, or in a word, authenticity."

In the minds of many, spirituality is necessarily connected with religion. Yet, in his authoritative book, *Spiritual Development—An Interdisciplinary Study* (1987), Helminiak argues otherwise. He points out the differences between psychology in the positivist viewpoint (which encompasses scientific aspects such as memory, cerebral functioning, cognition patterns) and psychology in the broader, philosophic viewpoint (which concerns the human need to be open, honest, and authentic). In essence, *spirituality is human authenticity*. Helminiak thus argues convincingly that the proper discipline to study spirituality is psychology within the philosophic viewpoint, rather than theology.

In his book, *Religion and the Human Sciences—An Approach Via Spirituality* (1998), Helminiak states: "Spirituality is a fully human affair. Spirituality is built into the human experience in the very makeup of the human mind" and emphasizes that, at its core, "spirituality is a human, not a theological matter; it pertains to social or human science and not to theological studies per se. Spirituality is an unavoidable consequence of being human." He clarifies further:

"Human spirit is understood to be one dimension of the human mind. To this extent the study of spirituality can be conceived as a specialized form of psychology, which has always been defined in the past as the study of the human mind. So on this understanding, far from being a theological matter, spirituality is primordially a specialization in psychology. It is that specialization that results when one studies the human being with explicit attention to the open-ended unfolding of the human spirit, which is specified by the inherent human requirement of authenticity."

Bearing in mind the earlier presentations on PRH, Hoffman's Quadrinity, and the two-tier model of the human person, we can conclude that spirituality, in Helminiak's understanding, really involves connecting the intellect, feelings, and body to the deepest inner core. *Thus, personal growth as promoted by sources presented earlier in the chapter is really a growth in spirituality.*

Religion, as an identified, prescribed relationship between an authentic human person and a revealed Creator Deity, certainly promotes human spirituality and may offer prescriptions for its development. Religion makes use of spirituality in a specific, transcendent perspective. However, one can be spiritual without being religious. One can even be religious, but not deeply spiritual, loyally carrying out prescribed requirements of a faith, but mostly from the limited awareness of intellect, emotions, and body. Lastly, a person can be spiritual and also religious by living with the intellect, emotions, and body, significantly connected to the inner core energy (being), and by recognizing and relating to this core energy in a revealed, transcendent context.

The key in spiritual, integrated human life is a personal, inner experience of one's deepest, authentic core energy. It has been noted that before the Industrial Revolution, human persons were indeed more integrated and more naturally inclined to wholistic living. With the age of industrial automation, particularly in the Western world, human life focus shifted significantly toward the intellect and its ability to control, and away from recognition of one true deep self. At times, what had earlier been spiritual *experiences* now became more of spiritual *understandings* and *principles.* For several centuries, this mode of operation was adequately energizing and progressive for human societies. Considerable inner security was derived from understanding and control.

Today, however, we are continually expanding our capacity for intellectual creativity and technical innovation. However, in many instances, we are losing control over our mode of life and thus losing inner security. Significantly estranged from our deepest inner source of fulfillment and security, unable to dominate our creation, we are at a loss. We need stability, motivation, resiliency, and potential for creativity, but past sources for such strengths (largely rooted in understanding and control) are leaving us dry. Intellectual principles for effective living, by themselves, are often proving operationally ineffective. Thus, we are beginning to search for vivifying inner experience from a dimension different than analysis and control.

Already over 25 years ago, physicians Meyer Freedman and Ray H. Rosenman, in a book on cardiac health (1974), remarked on the "rapid disappearance of myth, ritual, and tradition from the lives of most of us" in favor of nearly exclusive reliance on mechanization, automation, and bureaucracy. They noted "how strangely sad we have become as we search without hope for the color, the glory, and the grandeur we think life should sometimes shed upon us." Freedman and Rosenman also soberly declared:

"This is the first time in the experience of man on earth that a large group of individuals is attempting to live in so absolute a spiritual void." (We note that these astute, prophetic remarks were made before the time of the PC and Internet.)

The rituals and tradition to which the physicians refer are not to be seen as mere diversions, but as occasions for authentic inner openness, for an active connection to the deep, experimental, spiritual inner core energy which is available to all gratuitously and which is always waiting to be accessed and expressed.

In the words of Zukav and Francis, presented earlier, "Spiritual growth is now replacing survival as the central objective of the human experience." We may indeed be starting to move onto a different experiential path from the one initiated by the Industrial Revolution.

A focus on spirituality is appearing in business literature. A relatively recent article in the "Business Watch" section of the Canadian *MacLean's* magazine was entitled, "A Spiritual Link in the Workplace." Author Peter C. Newman refers to Martin Rutte, a consultant who is staking his career on the novel notion that the profit motive and spirituality can mix. Rutte is quoted as saying: "By spiritual values, I mean those values that lie at the core of our humanity, that come from our highest, deepest self…In my experience I've found that…we tap a powerful source of deep fulfillment and creativity." Rutte, billing himself as a "vision coach," points out:

"Spirituality is an experience…it's a connection with the living light…We're in a paradigm shift. There will emerge…new ways of work. Environmental degradation and lack of fulfillment are coming to an end. Respect and calling forth of people's individual gifts—spirituality—that's what's coming in."

The book, *Inner Excellence—Spiritual Perspective of Life-Driven Business,* by Carol Orsborn (1992), echoes Rutte's beliefs. The author declares, "The competitive edge in the coming decades will be held by those individuals and companies who can tap into new, life-driven sources of inspiration, creativity, and vitality."

She then proposes seven principles of life-driven {spirit-driven} business, and asserts: "We are pioneers poised on the brink of yet another frontier: the human spirit." She also points out that "the challenge of our generation of leadership is to cut through the tyranny of the illusion of control and expand our definition of success to include spiritual qualities and life experiences."

We are seeing a definite and increasing fraction of people looking inward for personal fulfillment and business success. In the last two decades, an increasing number of educated (and courageous) professionals, among them no doubt growth-prone IT workers, have been striving to go beyond analysis, control, and intellectual principles in order to re-connect with inner, experiential realities. They have referred to ancient, religious sources or to secular growth programs. Individuals have been seeking and often finding deeper experiential awareness in Buddhism, Taoism, Confucianism, Kabbalic Judaism, Eastern Christianity (with its sensorial emphasis and its "theology of being"), or various meditation programs.

Thus, it may indeed be worthwhile to ask *whether there is a common thread in all these approaches,* one that applies to every human being and that can offer on a recurring everyday basis, authentic psychological power, particularly to often over-extended, demoralized, and stressed IS professionals. If so, the *context and interpretation* of such an inner power (religious, philosophic, etc.) can rightly be left to personal choice, but the *development of mechanisms* for connecting with and applying such a power at work can indeed be a feasible goal for all those attempting to develop psychological awareness, inner robustness, and emotional intelligence in IT workers.

It is proposed here that the two-tier model of the human person provides the essential practical vision for *understanding* the mechanism of spiritual power, while PRH, Hoffman, and similar programs provide the means to *operationalize* these mechanisms on a daily basis. Helminiak, on the other hand, provides a very sound and innovative *theoretical foundation* for respectability and relevance of exploring spiritual (deep inner self) realities in a quest for wholeness, authenticity, and empowerment, regardless of a religious or philosophic frame of reference.

Daniel Helminiak also foresees development of a separate academic discipline of "human spirituality," distinct from psychology and theology, stating that:

"...a science of spirituality would indicate what is humanly wholesome and how far...human potential might be attained...a spiritual technology would emerge and it could be shared universally. Just as technologies for production, travel, communication, and so on {notably information} have transformed

the face of modern living around the world, so a spiritual technology might transform postmodern life. "

Such ideas, backed up by compelling academic argument, have prompted the assessment of Helminiak's book on religion and the human sciences as a "brilliant, scholarly, and breakthrough book" framing a vision of a scientific revolution in spirituality and the human sciences. Helminiak proposes that a spirituality worthy of all humanity would actually *inform secular culture.* In such a well-grounded vision lies considerable potential, both for theoretical research and for very pragmatic applications for the work culture of IS professionals.

Exposure to respectable, visionary academic insights on the integration of intellect, emotions, and spirit may be of particular interest to academics who have, in their research, already applied some aspects of psychology to the IT field. There are numerous MIS and interdisciplinary researchers who may have considered specific psychological factors in, say, research on artificial intelligence and human-computer interaction. Some of these researches may become the pioneers in the founding of the earlier suggested academic discipline of *Psycho-informatics.* This field would concern itself, theoretically and empirically, with understanding how different psychological mechanisms within and among IT workers and users affect the development and implementation efforts for computer-based information systems.

Having introduced the reality of the deepest inner (spiritual) self and having presented arguments for its serious consideration as a source of empowerment for IT workers, we now shift focus to the *mechanics* of developing a greater inner awareness and more direct understanding of how such an awareness can influence daily personal and work life.

INNER GROWTH - DEVELOPING INNER ACCESS

The central point of this chapter is that deep within each human person there exists a dynamic vivifying, motivating, and purely positive reality, which can have an immensely empowering impact on the daily functioning of our intellect, our emotions, and even our bodies. The second point is that most people (perhaps as high as 95% of the population, at least in the Western world) are significantly unaware that they possess such inner potential. While some people do maintain at least a weak and fluctuating connection to this genuine dynamism (although they may not be explicitly aware of this), many individuals have never experienced this deepest self to any appreciable degree (such persons may even argue vehemently against the existence of such a "non-scientific" inner reality). Thirdly, this deep inner core is now being recognized and actively promoted in various personal development programs by people who have undergone radical transformation in their consciousness, capacity, and personal fulfillment. Moreover, some programs have developed working models of how this genuinely energizing reality acts on the human person and how this inner psychological energy source can be systematically discovered and accessed. Fourthly, it is proposed here that awareness of and inner connectedness to the deep inner core energy can provide a "quantum leap" in an IT professional's capacity to handle rapid and continuing change, uncertainty, conflicting demands, time pressures, and other recurring stresses. Such psychological integration can also enable harmonious work relations and significantly enhance creativity and motivation.

Because this deepest inner reality (which can be referred to under various names pointed out earlier, including "one's spiritual self") is less widely identified and currently accepted than the intellect, emotions, and body, it has taken a number of pages to introduce it and to describe it with at least a minimal degree of thoroughness. At this point, the curious (but, perhaps, still skeptical) reader will likely thirst for the practical side: "Assuming I carry such an empowering energy deep within me, how do I access it and how will it affect my psychological experiences in my personal and professional life?" This and the next few sections attempt to address exactly these issues.

Impediments to Inner Growth

The word "growth" can take on several specific meanings. It can refer to professional growth in competence (as highlighted in the Couger and Zawacki study relating to system professionals), to developing new intellectual insights (as in discovering one's Myers-Briggs type), to adopting new life principles (as in looking for win-win situations), to developing new functional abilities (as in public speaking through Toastmasters), and so on. However, such growth, professional or personal, involves the intellect, some emotion, and use of the will. Such growth, laudable as it may be, proceeds at the one-tier or *horizontal* level (refer to Figure 3).

In these sections, we refer to "inner growth" as being mostly the *vertical*, that is, involving a significant shift in consciousness from a one-tier to a two-tier reality (from Figure 3 to Figure 4). Such growth involves getting in touch with one's deepest, authentic self (core energy/spirit). A natural question that arises at this point is: "If the deepest self is central to true empowerment and fulfillment, why are so many of us only dimly aware of it (if at all)?" Without attempting *the* authoritative answer, the following are seen as main, contributing factors:

1. The human being is born not fully integrated/connected/together—many have referred to this reality as "the human condition."

2. For the deep inner self to emerge to our consciousness, our developing psychological awareness needs to be "watered like a plant"—our unique traits and unique, undeniable dignity needs to be adequately seen, heard, trusted, recognized in individuality and worth, and safe to open without judgment, by persons centrally important to us; however, such persons can only provide attention from the degree of their own inner awareness, or lack thereof. To the degree that the above growth needs are met, the deepest inner self emerges (gradually) and becomes accessible to the intellect, emotions, and body—to the degree that the above growth needs are not met adequately, emotional blockages (many subconscious) set in, and one remains ignorant of one's deepest stabilizing strength and essential identity, possibly for an entire lifetime.

3. For the emotional growth needs that are not satisfied adequately, we develop compensating mechanisms (quite possible in one or more of the Enneagram ways)—while some such mechanisms may indeed be peripherally very functional, others may be more obviously addictive and self-sabotaging. We then react to other unfavorable circumstances throughout life through these mechanisms and may perpetuate further inner blockage from our authentic self. For example, a person's deep, essential intrinsic worth was not adequately mirrored during developmental years—she became a perfectionist for emotional security; now, rapid changes and unrealistic deadlines make it impossible for her to feel that she is doing her job near-perfectly, so she maintains an anger at others and disgust with herself, further distancing from her deep positive core.

4. Our culture, as a whole, is mostly not attuned to the reality of the deep inner self, on a daily basis; often achievement, possessions, prestige, appearance, etc., are valued in the absolute—"happiness" is derived from social acceptability or a string of temporary emotional "highs." The search for personal wholeness and solidity from within, apart from peripheral factors, is hardly promoted or encouraged. A large factor of

(at least) the Western world's population could benefit significantly from inner growth. However, such growth, although not hard to understand, is not easy to achieve. Inner emotional blockages, buried unfavorable experiences, and limiting compensating mechanisms need to be identified, accepted, and experientially expressed. For many, the yet unexperienced benefit (positive, deep awareness, which accrues slowly) is not worth the cost of abandoning old patterns and addictions through often uncomfortable inner evacuations.

Suggested Steps for Inner Growth

For those who are courageous, trusting, and motivated, there are specific, suggested steps for moving experientially from a one-tier to a two-tier consciousness, keeping in mind that the growth process will indeed vary from person to person. Firstly, one will likely want to investigate this "growth-in-being" more, to satisfy one's intellect. One can begin by reading personal growth books that promote the deepest inner self, such as those referenced earlier. One may wish to note common threads among these sources and compare them with one's own awareness at that point. One may also browse relevant websites for authentic testimonies. The site of the Hoffman Institute may have particular impact, since this program has had a considerable number of high-profile clients. As well, one may attend a growth workshop that promotes the two-tier model fairly explicitly.

At some point, one has to decide whether any of the material examined "speaks" to his life history and current life/work situation. If so, finding someone trustworthy who has done considerable growth work can help to assess different programs and to solidify one's strategy for further personal growth. In fact, stages in such a strategy can parallel the main phases of the system development life cycle. One must beware, however, of mistaking intellectual investigation of inner growth for the growth process itself (reading and talking about the violin is not the same as playing it).

After an introductory "feasibility investigation," if one has agreed in principle to starting inner growth, it will be helpful to keep two realities is the forefront of one's consciousness:

1.   The two-tier model (Figure 4): The aim of the inner growth is not to *understand* more or to modify consciously one's behaviors and attitudes, but to *open up* to an already existing, life-giving inner reality on another deeper level.

To this end, it is very useful to search for experiences in one's life that may have already opened at least a partial connection to the deep, inner self; experiences which have led one to deeply felt, unwavering certitudes, undeniable and undoubtedly true convictions about certain aspects of oneself.

If one has several such inner certitudes to work with (e.g., "I am a good listener" or "I am an excellent project manager"), then one would be wise to connect with them consciously on a regular basis, not so much to *think about* the occasion for certitude (math, listening, etc.), but to savor the inner *experience* of the certitude. It is very important to realize that a certitude rooted in the deep inner self will enable the experience of solidity and inner strength, but at the same time gratitude and humility before this talent as a "gift." Only such an experience is a channel to being. Simply feeling an "emotional high" for having achieved an exceptional grade, accompanied by pride and desire to dominate is a gratification at the "ego" level and is not a sign of opening to the deep inner core energy.

Having deeply felt certitudes about certain aspects of one's life is a valuable start in one's inner growth for it provides identifiable avenues through which the energy of the deepest self can rise to one's consciousness. However, at this stage the inner energy is (sometimes, and not necessarily at will) accessible only through discrete narrow channels linked to some exterior talent. (It is like breathing fresh air from outside through two or three narrow holes in the wall rather than fully inhaling the fresh air through a wide-open window). This limited accessibility can be easily choked off by unawareness and blockages that still exist in other major areas of one's life. The eventual goal of inner growth is to be able to experience this deep empowering and stabilizing energy, not through being sure, deep within, of some specific area, but through being deeply, inwardly convinced that one is, at the core, deeply good and infinitely valuable just because he/she exists, irrespective of his/her achievements, failures, or difficulties. This deep and real energy is always capable of expansion, providing a convincingly and increasingly felt knowledge of one's essential goodness and dignity. This power is indeed revitalizing in the face of outer difficulties since it comes fully and freely from within and is not dependent out outer circumstances (but its accessibility is dependent on an inward ability to open to it).

2. The phenomenon that, in addition to physical eyes and the eyes of the intellect, we have "inner eyes," the eyes leading to the profound inner conviction of the deepest self—this is the difference between the eyes of knowledge and the eyes of awareness. In the book, *Awareness* (1990),

the late Anthony deMello expands on the difference between *analysis* and *awareness*.

Consider an alcoholic who reads books and goes to all kind of workshops on addictions; yet, he continues drinking. His eyes of knowledge realize he has a problem, but his eyes of awareness are still blocked. The moment those inner eyes see that he is destroying himself, this profound realization and its acceptance opens him up to the power of the deepest self, and this core energy gives him the power to stop drinking as of that moment.

Emotional blockage puts blinders on our inner eyes, which were destined for truth. In growth work, we learn to see the truth of ourselves with the inner eyes of awareness. We need to "see" our strengths (the certitudes described earlier), and also we need to see (with no self-judgment) what blocks us from even greater strength.

For example, Cathy, a senior systems analyst, had excelled in high school, finished second in her graduating college class, majoring in Applied Computer Science, and has had three job promotions in the last five years. However, this lady had an older sister who was indeed a scientific genius and finished high school at age 14. Cathy was always compared, explicitly or implicitly, with her sister to whom she could not measure up. Cathy's intellectual eyes (for factual, analyzed truth) see and recognize her achievement and work successes, but her inner eyes of awareness (for deeply felt truth) have been "blindfolded." They are telling her, at a level much deeper than intellectual knowledge, that she has really achieved very little—too little to mention. When Cathy's inner eyes get to see the real truth and she accepts and feels it emotionally, an opening to the deep inner self will have occurred. This is a small-scale example of growth in the deepest self. As pointed out earlier, the goal is, eventually, to see with the inner eyes the undeniable, convincing truth of one's innate dignity and goodness, regardless of achievement, performance, and outward success. Acceptance of such a deep truth, at all levels, provides authentic power. Thus, Wayne Dyer proclaims that, as we get attuned to the "awakened life," higher productivity, achievement, and success matter less and less, but happen more and more.

Analysis - Where Do I Need to Grow?

With a preliminary assessment of growth and the view towards connecting one's two tiers of existence as well as an appreciation of the inner eyes of felt truth, one may indeed be ready to begin his growth process. In systems life cycle terms, preliminary investigation of growth is complete and analysis of one's psychological life begins. Here, one can employ the eyes of the intellect supported, where possible, by the inner eyes. One looks at his emotional history, from the womb onwards, keeping in mind the basic needs for the deep inner self to surface to one's consciousness: to be seen (and not ignored), to be heard, to be trusted, to be recognized with dignity, and to be safe to open without judgment. One searches truthfully without idealizing or excusing real deprivations. One asks: In my development years, what did not happen that should have happened, and what did happen that should not have happened? How deep and genuine was any recognition that one received? Was it conditional on some external result? What were specific moments when, through one's inner eyes, one internalized at a level far deeper than that of intellectual assent, that one is indeed "good" (genuinely at something, e.g., study, sports, sociability [all peripheral], or good in essence, a human person with undeniable, unconditional intrinsic dignity [central]).

During such analysis, it is important to notice and name specific *feelings* that one had during individual episodes of in-depth recognition or deprivation. It is here, already during early development, that inner access paths from the first-tier to the second-tier, to the deepest inner self, have either begun to develop or were becoming blocked (or possibly a combination of both). One looks very truthfully at the role of parents, educators, and other important persons in one's development as to providing a climate for the deepest inner self (stabilizing core energy) to begin to emerge, so that the developing mind, emotions, and body could have access to it. One can also look at the effect of one's culture and environment on one's inner awareness and potential for a deeply rooted, energizing life. (It is postulated that each culture [various ethnic cultures included] has both elements that facilitate deep inner openings and elements, not yet mature, that may induce inner blockage.)

What effect did peer pressure have in shutting down the emergence of specific, unique psychological traits? To what cultural (or counter-cultural) norms did one conform at the expense of an authentic personal identity? The occasions, or ongoing situations, where one's true, deep, and authentic self was not allowed to emerge and exist, for whatever the reason, have created the "wedges" in the two-tier model. They remain today in one's personal and professional life, preventing significant access to one's most valued inner resource.

After identifying, sometimes painstakingly and often surprisingly, that there indeed are (as there are in about 95% of people) notable inner barriers to deep inner energy, fullness, and tranquility (the kind of stabilizing, vivifying energy that would still be undeniably present even if everything else in one's life could not be counted on), the next step in this kind of self-analysis is to identify, again very truthfully, the *compensatory mechanisms* that one has developed to derive some psychological energy for daily living and personal progress. Initially such mechanisms (e.g., rapidly progressing in academics, creating a new technologically oriented enterprise, etc.) may not look to be compensatory (the compensation may have largely been subconscious), but if one has significant emotional attachment to their involvements, *without which life would seem significantly less meaningful and the person would feel significantly less valuable*, there may have been a hidden motive and need for such intense involvement. One will now see that life energy derived in such ways, although propelling one forward and perhaps, indeed, providing a significant degree of satisfaction, was a peripheral (and not deepest, most potent) life energy which was sought after, since the deepest central life energy was not adequately available for access.

Compensatory mechanisms may have been quite noble, acceptable and might have indeed felt satisfying; alternatively, they may have been more obviously destructive. In either case, however, an emotional dependency was formed, an insatiable *need* for something on the outside to feed an inner craving that arose from an unawareness of and inaccessibility to one's deepest inner core. For example, an athlete, student, or musician *needs* to perform at an excellent level to feel acceptable and valuable. A manager *needs* to be in control to feel "OK." A volunteer strongly and repeatedly *needs* to see how much her work is making a difference in the lives of others. The key word here is *needs*. Where honorable activity is used (albeit subconsciously) as a compensation resulting from unawareness of one's essential, unconditional goodness, dignity, and life energy (emanating from the inner core), the need for success becomes acute and its frustration causes disproportional emotional disruptions. Thus, one can identify one's main emotional involvements, starting in one's earlier years, and one can assess if there indeed may have been significant compensation (as is likely in the majority of the population). For example, did one become a volunteer in a nursing home *because* one was essentially contented (having been adequately seen, heard, trusted, etc., as a growing person) and working to express their contentment in assisting others, or did one become a volunteer *in order to* gain contentment from the interactions, which were otherwise fairly minimal. (This may not be an either/or situation, but, what was the predominant reason?) Similarly, did one program computer games largely as an expression of one's essential inner life energy that was already present regardless of one's choice to program, or did one write the programs mainly in order to gain the majority of one's life energy and self-worth from this activity? In such insights, we see the root of what may be termed as "noble addictions."

It can sometimes be that an initially compensatory activity (especially if it follows a genuine talent and is adequately received) may, in time, become an occasion to some real opening to one's deep inner self. In many cases, however, compensatory activities remain as such throughout a lifetime, at least until this type of growth is undertaken. In *Working with Emotional Intelligence* (1998), Daniel Goleman presents, from Robert Kaplan's research, common, costly blind spots of 42 business executives: blind ambition, unrealistic goals, relentless striving, driving of others, hunger for dominance, need for recognition, need to seem perfect, and preoccupation with appearances and image. One may indeed be totally unaware of the compensatory nature of such compensatory drives. Not having significantly "tasted" the immensely vivifying and empowering energy of the deepest inner self, one is not aware that there is anything for which to compensate. This can be a major reason for initial resistance to, and perhaps a cynical assessment of the inner growth process, particularly among largely analytically/cognitively oriented (through profession and culture) IT workers.

With an intellectual understanding and emotional acceptance of the reality of one's blockages from genuine inner connectedness, as well as of one's predominant compensating mechanisms, the analysis stage largely comes to an end. It is worth noting that while analysis may be the preface to inner growth, it is not growth's major catalyst. *One cannot think or will his way into his deep inner being.*

Development in Inner Growth

Now comes the time to plan and carry out (design and develop, as in RAD) one's personal growth programs. Firstly, one must see (with the inner eyes) that this is the right time. Forced growth, fuelled by willpower, may significantly shut down even further the connection between the human tier and the being tier. The goal of the growth process is, having identified emotional barriers to wholeness, to release them gradually, from one's entire system. This will at some point inevitably involve deeper feelings that may not always be pleasant.

To this end, a *growth facilitator* who fits well with one's own inner "vibrations" can be very helpful. It is important to note that there is a difference between a *counselor* and a *growth facilitator*. A counselor, more often than not, will receive a client mostly from his/her intellect and deeper feelings, whereas a true inner growth facilitator receives a person from his/her own inner being, to which he/she has frequent access. Such a level of inner presence and communication eventually initiates the client's connecting to his own inner being, however limited such a connection may at first be. In such an atmosphere of receptivity, the client will often be spontaneously motivated not just to *talk about* (intellectually), but to *express* (with deep emotion) incidents of inner blockage in one's life history.

For example, Cathy, the systems analyst with the genius sister, may experience, maybe for the first time in her life, the depth, intensity, compassion, respect, support, and most of all *presence* of the inwardly rooted facilitator. Such profound receptivity may indeed prompt Cathy to re-experience spontaneously different feelings from specific incidents in her past where she was made to feel inferior to her sister. She will let her whole self experience those long-time blocked emotions, perhaps breathing or crying them out, allowing her body to vibrate, lash out, tighten, or relax, whatever is her authentic expression at the moment. While she is doing all this, Cathy is explicitly aware of the deep receptiveness, empathy, and truthful support of her authentically rooted and attentively present facilitator. She takes in the deep energy from her, and this stimulates her awakened, albeit dimly, deep inner self, permitting further evacuation of blocked emotions. At the end of the session, Cathy feels a deeply satisfying inner opening and begins to experience a sense of unconditional worthiness and aliveness that she has never felt. She thinks to herself: "All the significant joys of my life—graduation, my first job, my marriage and honeymoon, my trip to Switzerland, and my merit award at work—simply pale as compared to this, what I have uncovered deep within."

For evacuating such as Cathy's to be effective in connecting to the deep inner power, considerable auxiliary support work may be necessary. Here, approaches will differ for different people (maybe based on personality type?). Daily meditation (repetition of a "mantra") can quiet the intellect and allow space for the inner core energy to surface to consciousness (recall the remark from Dr. Bernie Siegel). Journaling allows one to realize, at a level deeper than the intellect, what actual situations in one's life history have not allowed adequate life energy to surface. Going into nature, alone, without analytic thinking, and without expectation, allows life at the sensorial level. Since the feelings and senses lead one closer to the energizing core than the intellect, stimulation by the presence of nature can be a valuable growth aid. Artwork, particularly when it is spontaneous, may again allow one to bypass the intellect and begin expressing what has been stored deep within.

Recurring dreams (perhaps about a stressful systems project) can provide clues as to what experience is blocked, often in the subconscious and awaiting appropriate processing and release. Moving music can also be very influencing, releasing blockages and enabling deeper connection. Recently, music therapy professor Teresa Lesiuk, in her doctoral research, found that music listening contributed significantly to mood change, stress reduction, and increased focus on work for information systems developers (2003). Films with deep emotional content may elicit emotional responses, which point to necessary inner release. For those with a religious orientation, various forms of prayer can be a catalyst for inner healing (and, conversely, inner unblocking can give more depth and experiential meaning to prayer and faith).

Various health aids can also be useful. It is important for a person in inner growth to notice and attend to physical, often chronic, and recurring pain with a psychological content (e.g., head or chest tightness, stomach uneasiness, bowel spasms, etc.). Reiki and other forms of "bodywork" can dislodge repressed blockages, enabling their release at the appropriate time. Yoga and tai chi move the inner "life force" that may have been lodged in specific bodily locations. *Homeo*pathic remedies (energy medicine) exist that address specific emotional states. Such remedies (which are distinct from other *naturo*pathic medications) can, in a few moments, increase stamina, allay fear, and dispel depression, since they affect the person's energy field directly without need for physiological assimilation. (It must be noted, though, that homeopathics are particular to the client, not the symptoms—thus two programmers may be using entirely different remedies to alleviate headaches with identical symptoms after an unexpected system crash.) Also, homeopathics have been known to relieve chronic conditions such as migraines with only several treatments. Nutritional balancing can also be influential in helping blocked emotions. In his book, *Your Body Doesn't Lie* (1994), psychiatrist John Diamond points out simple kinesiological tests to determine if a particular food substance would increase or decrease a person's energy at that given moment. As well, formations of "growth groups," physical and/or virtual, among IT workers, can provide a very welcome forum for sharing knowledge as to emotional, physical, and intellectual integration. Such groups can also be a welcome forum for sharing deep feelings, especially feelings triggered by specific work events.

Parts of the Person in Inner Growth

Having, now, a basic orientation to the inner growth process (the purpose of which is to connect to the deeply energizing inner being energy) and referring again to the two-tier model of the human person (Figure 4), we need to examine the role of each component of the person in doing inner growth. *We cannot use the same approach ("brute force"/willpower) to build inner awareness as we may have used to build our careers.* A new paradigm must be recognized and appropriated.

In inner growth, the *intellect*'s role is very secondary. Initially, it may be quite useful in identifying causes for inner unawareness and compensation mechanisms. From then on, it takes a "back seat," observing, making rational sense of what is going on, but in no way directing the process. The role of *feelings* is paramount. Profound emotions relating to not having been adequately allowed deep inner life need to be expressed at the right time (where there is enough positive energy within and enough time and space) and in the right environment (when the expression is received at an adequate level of depth). It is noteworthy, here, that most addictions, including addiction to work, success, or control, result from profound, yet unexpressed feelings. Yet it is not enough to *express* feelings; the body and its senses must also be involved. The muscles and nerves must be allowed to finally "let go" of what they had been retaining for so long. If we accept Dr. Siegel's assertion that "feelings are chemical," a bodily release of blocked feelings will indeed have profound effect at the biochemical level, prompting much more effective nutritional assimilation (it has, in fact, been suggested that specific feelings, such as anger, grief, etc., will block the assimilation of specific foods such as wheat, sugar, etc.). As well, such emotional release should affect the health of nerve and brain cells, and increase the level of physical stamina and vitality.

We can thus recognize that inner growth work is indeed wholistic; the involvement of the intellect is moderate at best. This, however, should not be an excuse for predominantly thinking types (STs or NTs) to dismiss inner growth. Thinkers, too, have feelings, many of them blocked and awaiting expression for significant psychological empowerment. It is noteworthy, thus, particularly for IT workers, that in recognizing the inherent interconnectedness of all parts of the human being, mostly non-intellectual work is prescribed so that resulting, unblocked energy can give new tone and power to the intellect. Some will find such a realization quite radical.

Characteristics of the Growth Process

There are other characteristics of the growth process worth noting. Firstly, it is non-linear and largely non-controllable. We move back and forth in awareness of both our blockages and our already emerged, deep true self. Also, the emergence of the deep inner self cannot be anticipated until it happens. The inner self is mostly eclipsed until it surfaces from the subconscious during the growth process. When it happens, one experiences profoundly vivifying and new awareness; until it happens, one remains largely in the dark. The deep self emerges gradually but gratuitously—it cannot be "pulled up" by effort. Where effort is needed is in comprehensive evacuating of blockages so as to give this gratuitous reality enough space within our consciousness; this harmonizing, stabilizing life energy will then automatically fill the space. The comprehensive evacuating can comprise several aspects. In his book, *ONE* (1977), computer scientist Orest Bedrij prescribes his "breakthrough formula" to consist of purity of heart, love, lack of compulsive attachment, and stillness of the mind. He warns that attempting stillness (e.g., through meditation) without clearing up internal pollution (repressed negative emotion) "can be equated to boiling a sewer that has not been cleaned." He then quotes another teacher, Edward Carpenter:

"Of all the hard facts of science, I know of none more solid and fundamental than the fact that if you inhibit thought (and persevere), you come at a region of consciousness below or behind thought and a realization of an altogether vaster self than that to which we are accustomed...So great, so splendid is this experience, that it may be said all...questions and doubts fall away in face of it."

This deepest self emerges, for the novice, gradually. First, one has but an inkling of a deeper presence within, then one gets recurring glimpses, and after significant inner growth, one gets a fairly recurring, fuller view. Finally, it would be idealistic to assume that a person can remain perpetually connected to one's inner being. The "human condition" is such that one will be "pulled out" of it again and again, but, in time, reconnection will be more natural and effortless.

As well, the intensity of our awareness and experience of our deepest inner self and its effect on our intellect, body, and emotions is capable of ongoing expansion. After considerable inner growth, a person will be able, through his inner eyes, to see himself (the true self), to hear himself, to trust herself, to value herself, and to feel indeed safe with himself (a sense of solitude rather than loneliness), both in his/her deep essence and in peripheral talents. Moreover, an astute, connected person will be able to recognize at any moment which of the four components of the person in the two-tier model is "in charge" (which, by analogy, is the "lead horse" in a team of four horses). She will be able to answer to herself at any point in time, ***"From where within me am I mostly experiencing this situation?"*** Indeed, she might react to a given situation in her professional or personal life very differently if she is being guided *mainly* by her intellect (principles, norms, ought-tos), her feelings and sentiments, or her deepest inner center. Such awareness, in itself, is of unquestionable value, and can readily be applied to the IT work environment.

However, while inner integration does promise a much deeper psychological energization and robustness, it cannot promise utopia. Human difficulties, accidents, or tragedies will not necessarily escape a considerably integrated person. It is possible, though, to live disappointment, grief, or frustration at one level while still remaining firm, robust, and essentially peaceful on another level to which one now has access. With growth, the impact of negative events in the psyche will not be devastating, and recovery will occur more quickly and definitively.

Growth Work and Psychotherapy

At this point, a clarification is in order regarding growth work and psychotherapy. Growth work guidance belongs to educational systems for personal psychological development. Its goal is to enable a person to know oneself better according to the two-tier model and to become significantly more rooted in one's essence with an unwavering conviction of one's essential goodness and dignity that is not dependent on external circumstances. Psychotherapy belongs to psychology/psychiatry, with the main aim being the healing of psychological traumas. Inner growth programs are aimed more at people who are operating functionally but sub-optimally on a daily basis and who sense a need to become more than they currently are. Psychotherapy programs are often aimed at persons who may be functioning only marginally and need definite assistance in order to function "normally." However, the two approaches do have some points in common—for example, work on the dimension of inner healing (unblocking).

The above exposition of personal, inner growth was intended to be introductory and descriptive rather than didactic and prescriptive. It, hopefully, has offered a curious reader insight on *how* a person "gets into" that vivifying psychological zone that is reported to offer such valuable benefits. Now, it is indeed worthwhile to examine how one who is sufficiently connected to his own depth might function in a work environment, notably an intellectual one such as IT.

FUNCTIONING OF AN INTEGRATED PERSON
Existing on Two Levels

Once a person reaches a significant level of integration—that is of connection, experientially, of one's intellect, emotion, and body to the deepest inner self—one's functioning and one's self concept will eventually change. How so?

In an integrated life, we exist on two levels (tiers). Here we can envision two concentric circles. Our primary aim is *to be*. To realize this, we make use of the talents of being which emanate from an accessible deeper inner self (inner circle). Such talents include patience, humility, wonder, confidence, respect, gentleness, joy, peace, security, openness to truth, and capacity for in-depth love. These are, then, the *central talents*. They are gratuitous, but need to be discovered within the deepest inner self. Our auxiliary aim is *to do*, that is, to discover our peripheral talents (outer circle), and to connect them and subordinate them to the central talents. *Peripheral talents* include thinking skills, writing skills, communication skills, technical skills, athletic skills, musical talents, learning talents, personality dimensions, and so on. In an integrated person (living on two tiers), one's peripheral talents are used to express the essential, primary talents; here the intellect receives, structures, and coherently expresses what is grasped from the inner self. The feelings and senses vibrate the life in the inner self, and the body encases the life of the inner self. Such a life generates profound fulfillment and meaning.

One issue which is likely to change significantly with the inner integration is that of *personal identity*. The inner self just *is*; it needs no other labeling identity. This "*is*ness" is enough for profound self-confidence (a central talent). A person with such an awareness is not emotionally dependent on one's performance in specific roles for primary confidence, value, and meaning. In any required role, one is free to do one's best and can accept fully whatever that may be. However, when a person is unaware of one's essential (gratuitous) security, he/she is often compelled to perform to a certain standard to maintain self-esteem. In rapidly changing circumstances such as IT today, this is a major source of stress.

With growth work, one has a powerful experience of his/her essential identity (from connecting to the deep self within). One thus realizes, with the inner eyes, that *one is not what one does, one is not how one behaves, one is not simply his personality or his achievements*—one simply *is*, and this is enough for in-depth security and primary fulfillment. One also realizes that others are not their work, their shortcomings, and their irritations (these arise from their lack of connectedness); others are also only positive in the inner core, although they may not have yet discovered or actualized this reality. Regarding work, an integrated person works *because* she is fulfilled rather than working *in order* to become fulfilled. Indeed, there is an added level of fulfillment in working with genuine motives according to one's developed, peripheral talents; however, frustration in such work, for the integrated person, does not cause extreme inner disruption or an existential crisis.

We can recall that, in *The Heart of the Soul*, Zukav and Francis echo the above two-tier perspective of integration when they declare, "Authentic power is the alignment of your personality with your soul." As well, they address the issue of inner security, particularly in times of rapid change. "The old species explored the physical world and it created security by manipulating and controlling what it discovered…the new species creates security by looking inward to find the causes of insecurity and healing them. This is the path to authentic power." Is not this authentic power perhaps the prime "soft skill" desirable of IS professionals today? Is not this authentic power a root for "emotional intelligence"?

"I Think - Therefore, I Am"?

The work of system developers, programmers, data analysts, telecommunications specialists, and the like does require primarily the use of the intellect, at times backed up by intuition. What is the role of the deep inner self here? Much of the Western world (and likely many IT workers) has apparently adopted the mindset of the philosopher Descartes: "I think, therefore I am." However, neurologist Antonio Damasio, in his book, *Descartes' Error: Emotion, Reason, and the Human Brain* (1999), provides compelling evidence from a physiological perspective that the work of the intellect cannot be separated from other components of the human person. The two-tier model, in fact, connotes *"I am, therefore I think"* as the optimal mode of being and functioning. The intellect is thus intended to be a servant of the wisdom and insight of the deep inner self. Indeed, the intellect has all autonomy. Thus, *the intellect can reject, condemn, be indifferent to, doubt, impose its agenda on the deepest inner self, or rather it can cooperate with and eventually be at the service of this inner self.* This is what Joan Borysenko refers to as making the mind one's servant, rather than its master.

> The Intellect can ...
>
> ... *reject* ...
> ... *condemn* ...
> ... *be indifferent to* ...
> ... *doubt* ...
> ... *impose its agenda on* ...
>
> ... *cooperate with* ...
> ... *live at the service of* ...
>
> ... the Inner Being.

Such a perspective was exemplified in the work of U.S. mathematician and computer scientist Orest Bedrij, who worked out fundamental mathematical equations for parallel processing of long series of numbers with multiple integers—equations which were then patented by IBM. Bedrij subsequently became IBM's technical director at the California Institute of Technology Jet Propulsion Laboratory, and was responsible for the development and integration of a computer complex that controlled the first soft landing on the moon. To discover the equations for parallel processing, Bedrij prayed and fasted intensely for a long period of time, as was his wont in discovering new formulas. He ceased any attempt at deliberate calculation and sat motionless facing a blank wall where the equations eventually appeared. Bedrij has published a unique, profound, and fascinating book called *ONE* (1977). The book integrates physics, mathematics, biology, economics, and religion, pointing to the underlying oneness of existence and humanity. It has been hailed as a "brilliant and powerful volume—of the highest wisdom" and will be indeed insightful, particularly to NT types. Also, Daniel Helminiak exclaims, "Though one may prepare for insight by study and hard thought, insight always comes unexpectedly as a gift."

Thus, there can be a significant difference in process and in result, between thinking with a detached intellect (where many feelings and a strong connection to the deepest self are blocked) and thinking with an intellect that is at the service of the inner being to which it has access. The former is often rigid and directionless—it may be like continually viewing one's surroundings through powerful binoculars without guidance as to how, when, and why to use them.

A disconnected intellect will often produce anxiety and inner unrest. Yet many intellectually oriented workers (many in IT) living on a one-tier consciousness simply cannot stop analyzing (unless they engage in diversion). They are addicted to thinking, yet often unfulfilled by it. When their thoughts are "overloaded" or confused, when they lose the control they maintain for personal security, or when their thoughts initiate feeling threatened or frustrated, possibility for effective (and rapid) re-generation and refocusing are limited. They can try to change their thoughts, yet they can only try to by willpower, which may be hampered by blocked emotions. Indeed, in my survey of 250 North American IS professionals, 63% agreed, at a level of at least five out of seven, that they were finding it hard to stop thinking about the job at day's end; 44% said that work problems kept them awake at night. Many such people are undoubtedly living (consciously or otherwise) the Cartesian reality: "I think, therefore I am."

However, an IT worker who has consciously grown to a two-tiered awareness and who is able to connect quite often to his pure, positive inner self/core energy, has a very significant advantage in dealing with such "thought overload" (he is living: "I am, therefore I think"). With practice he can abandon, inwardly, the confused thoughts and resulting discomforting feelings, and open himself up to the deeply felt reality of his goodness and dignity. This is a real and powerful energy that can revitalize him in the face of significant outer difficulties, as this inner reality *is not dependent on outer circumstances.*

Personalities and the Deepest Self

With considerable focus as to personal psychological awareness having been placed on personality types, two main issues must be discussed relating personality to the deepest inner self. Firstly, *what is the relationship between these two dimensions of awareness?* Considering again the two-tier model of the human being, personality relates mostly to the first tier (intellect and intuition, feelings and bodily senses). We can recall the definition provided earlier: "Personality is a complex set of relatively stable behavioral and emotional characteristics of a person." However, in the two-tier understanding, personality type does not *define* a person; it explains, according to a coherent and empirically tested system, how certain classes of persons tend to function, where they tend to focus, how they assimilate what they perceive, and what underlying motives may be fuelling such functioning. Since most people tend to live on a one-tiered awareness, personality identification is easier to assimilate than the reality promoted in this chapter. The limitation, however, lies in seeing oneself, from the psychological perspective, as (nearly) entirely one's mode of functioning/relating. Although the phrases, "I am an ISTJ" and "She is a Nine," are commonly used, the intended understanding is really, "I function more as ISTJ than as other types" and "She behaves like a Nine type."

In the two-tier model, *personality is "informed" by the deepest inner self* (the second, deepest tier). In a well-integrated personal and professional life, the personality is connected to the deep self (core energy). Thoughts, feelings, senses, and intuition do not "gallop" on their own, but coherently express and vibrate the wonder, insight, and consciousness of the deep-but-now-fairly-accessible inner being. Using an auto mechanics analogy, the personality is like the transmission, while the inner core is the engine. In the words of Zukav and Francis, "Authentic power is the alignment of your personality with your soul."

As an example, consider two IS system developers, Tom and Cathy, both ISTJ (Introverted, Sensing, Thinking, and Judging) types on MBTI and Sixes (Loyalists) on the Enneagram. Such persons will be detail-oriented, wishing to "get it right." Tom and Cathy's personality awareness, as well as that of their supervisor, has placed them in IS positions for which they have a natural inclination. Yet, while each will approach IS tasks with a similar *style* and attitude, their inner emotional *experiences* may likely be quite different. Moreover, such differences in inner experience and, consequently, the availability of psychological energy may impact significantly worker productivity and effectiveness. (Development of research efforts in this area should certainly be encouraged.)

While both are aware of their personality types, Tom has little, if any, experiential sense of his deepest core self. Cathy, however, having done considerable inner growth work, is able, to at least an appreciable degree, to live from the deep inner space which echoes "no matter what happens, I AM." In a tight deadline situation where an expected major program error has been discovered, Tom will likely react in fear and anxiety; his loyalty to the project and its deadline must be maintained, yet this may not be possible. The disruption in the anticipated schedule may significantly "rock" his needed sense of control. As an introvert he will keep his escalating negative emotions hidden, i.e., largely repressed. He may likely not be able to "turn off" when leaving work and may have difficulty sleeping. Tom is close to his project—a loyal, capable, caring IS employee. However, he has difficulty being *emotionally free* of his work.

Cathy, on the other hand, has access to a deeply helpful inner resource. She, too, is initially shocked by the project disruption. She, too, is loyal and takes her assignments seriously. However, her personal self-worth, and therefore her stability and ability for regenerating psychological energy, do not at all come primarily from needing continual success-feedback from her work or from respect of her peers and supervisors. She acknowledges psychologically the difficult work reality, but adjusts to it readily. The deepest inner core energy, to which Cathy has access, is always positive and vitalizing. As mentioned before, it is present in everyone and is gratuitous (like oxygen); however, persons in growth must "dig a tunnel to it" by removing systematically and experientially inner blockages and unawareness.

Cathy's inner being provides stability and an overriding confidence. It provides her the ability to accept emotionally the truth of the situation. Comforted and strengthened from within, she does not feel persistent and debilitating fear, anxiety, or anger. She is, thus, able to be fully attentive to the situation at hand and to marshal all her ability, experience, and insight to try her best to solve the problem. Above all, while Cathy, being a Loyalist, very much *wants* to overcome this problem (and is inwardly free to do so) at the deepest level, she does not *need* success in this situation or even in this job to feel valuable, dignified, and energized. In this situation (of which there are many in today's IS climate), Tom *needs* to succeed, emotionally, and thus may overfocus; Cathy very much *would like* to succeed, and is able to maintain perspective. Tom reacts, while Cathy responds.

In this and similar work situations, the critical success factor for an IS employee may well be, not his/her technical expertise or even interpersonal skills, but *the inner power for deep emotional stability and resilience*. With the two-tier model, promoted by PRH, Hoffman, and numerous other programs, we have an understanding of and methodology for developing such inner strength. Quotations regarding the deepest inner self, some of which were presented earlier, often refer to "power" and "empowerment." This, of course, is not the "power" of the prevailing one-tier business culture, which uses willpower for control and expansion. This is an authentic power of a Higher Order, which respects, harmonizes, and stabilizes. It is not a power *over*, but a power *for.* In the face of developing efforts for personal integration with the deep power, in the face of considerable empirical evidence of the personal and professional consequences of such power, in the face of the hypothesis that, while personality awareness certainly has valuable insights it does not probe deeply enough into the human psychological reality, can the IS profession and MIS/interdisciplinary research afford to ignore this developing paradigm?

If we understand personality more as a way of functioning rather than an essential identity, we can appreciate that an opening of the personality to the deepest inner self can help to balance and round it out. In MBTI an introvert, having connected to her/his deep inner authentic self, may now be more prone to express it. A sensing (practical) person who may have (subconsciously) used emotional attachment to the practical and tangible for inner security, now being secured in her core, may naturally begin to appreciate hidden meanings and new possibilities of the intuitive. A predominately thinking person with little access to feelings will certainly be more comfortable in feeling deeply after the type of inner growth process earlier described. A judging (structured) person may be less rigid and more tolerant of ambiguity. Balanced personalities who are comfortable in accepting others' differences certainly help in the effective functioning of IS project teams.

It is also worth noting the subtle difference of relationships of the deepest inner core with the Myers-Briggs and Enneagram systems. Myers-Briggs, developed in the West, is more exclusively on a one-tier level. The only allusion to a deeper intrapersonal reality may lie in the description of the NF temperament. The Enneagram, on the other hand, with ancient roots in the East (which was not as directly affected by the Industrial Revolution, and resulting intellectualization) contains explicitly "directions for integration." While it does not identify the reality of the inner core explicitly, it does indicate, on the $1 \rightarrow 7 \rightarrow 5 \rightarrow 8 \rightarrow 2 \rightarrow 4$ subsystem, that satisfaction of deep personal needs occurs at a move towards 4. This may well be the strongest point of interest in and solid commitment to an inner growth program such as PRH or Hoffman's Quadrinity. On the $3 \rightarrow 6 \rightarrow 9$ subsystem, it can be hypothesized that most growth in the inner self would occur at the move towards 9, where external image or security in external dictates are surrendered in favor of personal serenity. Further research into this question may be in order for competent, motivated, and creative investigators.

How Different Personalities Might Address Inner Growth

While the first issue relating personality to growth-in-being is the functional relationship between the two, the second issue relates to *how different personalities/temperaments are likely to address the possibility of growth into the deepest inner self.* Although what is presented here falls mostly into only "reasonable speculation" and engages in unavoidable generalization, it may offer insight into pitfalls that different types may well avoid on their way to personal wholeness.

In terms of MBTI temperaments, the NF (intuitive feeler) is undoubtedly the most naturally predisposed to seeking for and being true to one's deepest inner self. This type has the deepest feelings with a central need for authenticity. (From my experience in giving a "wellness workshop" for IS professionals, it is by far the NFs who were most excited and appreciative on my module on "One's Inner Core".) *Thus it is the NFs in the IS field who will have a central role in developing psychological awareness among IS workers,* where they are allowed such a role. As the profession matures in emotional literacy, it is the NFs who will be viewed as a special type of an organizational resource, by virtue of their temperament. However, NFs will need to develop patience with other types regarding interest in and perceived immediate need for descending into another layer of inner life. Also NFs may need to modify their vocabulary when introducing broad audiences to the deepest inner self (e.g., many STJs and NTs may be more likely to resonate with the phrase "authentic power" rather than "spiritual journey"). As well, NFPs in particular may need to balance, in their verbal and written presentations, the open-ended, non-linear nature of inner growth with identifying tangible milestones and competencies arising during the growth process. The majority of Js in IT still expects results!

NTs (intuitive thinkers), organizational and technical visionaries, may be open to the *idea* of the deepest inner self, but will first want to *know* more about it, and to understand growth mechanisms. For this type, the volume, *Persons and Their Growth* (the anthropological and psychological foundation of PRH education) (1997), a collective work realized by PRH International, will definitely provide unique food for thought. *No One Is to Blame*, by Bob Hoffman, may also prove to be enlightening, as may elements of the newsletter, "The Light News," published and distributed by the Hoffman Institute. For further foundational insight, *On Being a Person*, by Carl Rogers, will likely be valuable. For those interested in the theoretical foundation of human spirituality (and who are prepared for considerable intellectual depth and rigor), *Spiritual Development* and *Religion and The Human Sciences*, both by Daniel Helminiak, should indeed make a significant impact. Of course, Bedrij's *ONE* should be a work of impact, particularly for broad-minded, adventurous NTs.

The pitfall for NTs regarding growth-in-being is confusing *understanding* and *believing* in growth with actually *doing* and *achieving* growth. Excitement about personal intellectual breakthroughs regarding human existence and functioning, authentic as it may be, is not be enough to carve a permanent connection to one's deepest inner self. The role of feelings (yes, even for NTs) in the unblocking and integration process is paramount. The deep inner self is pure, positive life force-energy; it is not the same as feelings. It can be *expressed* through various human dimensions, including intellect and feelings; however, feelings are involved in reaching down to the core and "connecting it" up to the intellect where NTs are so comfortable.

SPs (sensing perceptives) are practical people with a bent for variability and spontaneity. Such people, in general, may have difficulty with conceptual presentations on the inner self (even the one in this chapter). They may also lack a structured, linear approach in identifying inner blockages systematically. Also the feelers among SPs may be more comfortable with sensory-related emotion than, initially at least, deeper, sometimes uncomfortable feeling. SPs may confuse warm feelings of kindness with the deeper energy of the inner being. However, this temperament would likely respond to authentic relatedness, perhaps from an appropriate growth facilitator. They may also appreciate the use of music, art, and even film to generate deeper inner connection.

SJs (sensing judgers) are practical, structured, and detail-oriented. They predominate IT work. From my experience from workshops and from teaching MBA students, SJs, at least initially, can be quite confused by a promotion of one's deepest, essential reality and core energy. Their lives are often influenced significantly by logic, structure, tangibility, and principles. For example, they may be more at home with spiritual *principles* rather than deep spiritual *experiences*. For many SJs, inner core growth may not be the best first step towards psychological integration (as it may be for NFs). Also, SJs' preference for traditional life values and approaches may distance them from novel, yet promising efforts. However, inner, second-tier growth may be initiated for SJs through detailed worksheets inviting persons to name their feelings in certain life situations. At that point, SJs' focus on detail may allow them to itemize "what happened that shouldn't have and what did not happen that should have." A guided, structured program would eventually move such types into the necessary feelings needed to evacuate blocked psychological energy. Also, since sensing is central to the SJ, exercises shifting the focus from the intellect to the senses may promote a consciousness of other experiential dimensions of their person. Considerable research and then also development would indeed be in order here. Yet the gratuitously available profound life force is as intended for full accessibility to SJs as to any other type.

A worthy goal of introducing psychological awareness to IS professionals is to relate and integrate *many* inner dimensions of the human person, especially those that have proven to affect significantly—by one's intrapersonal harmony—interpersonal effectiveness, and the quality and productivity of work. The approach in this book, thus far, has been to motivate and hopefully initiate such a path.

AN EXPLANATION OF "EMOTIONAL INTELLIGENCE"

For a long time, the Western world has considered IQ, the ability for cognitive reasoning, as the main (if not only) determinant of a person's "intelligence." In IT, it was indeed very often the case that the person whose programs were most efficient and had the fewest errors, would be considered as the prime candidate for promotion. However, today, in the "post-modern" era, the development of information systems is recognized as a socio-technical field. Nearly a decade ago, an article in *Computerworld* called for "emotional literacy" among IS developers. Since then, a number of articles have appeared calling for development of "soft skills" and lamenting that IS workers are often emotionally weak. Such articles often call for another type of intelligence in addition to superior cognitive, logical skills.

Harvard psychologist and educator Howard Gardner had, in the 1980s and beyond, promoted his theory of multiple intelligences. He maintained that in addition to logical/cognitive intelligence (as measured by IQ), there exists linguistic intelligence, and also naturalist, interpersonal, intrapersonal, spatial, musical, and bodily kinesthetic intelligences. Following this type of perspective, Daniel Goleman, with a PhD in Psychology who formerly taught at Harvard, in his 1995 international bestseller, *Emotional Intelligence*, focused widespread attention on the reality that what matters regarding one's ability to succeed in work and life is often more than IQ. He has, however, stated that EI is more than one thing and includes: knowing what you're feeling, using that knowledge to make decisions, being able to manage distressing moods, maintaining hope in the face of setbacks, having empathy, and being able to get along with people. In his book, Goleman sets out on a quest "to understand what it means—and how—to bring intelligence to emotion." He also declares that "the market forces that are reshaping the work life are putting an ***unprecedented pressure on emotional intelligence for on-the-job success***" [bold added]. Goleman shows that brain circuitry—neurological exposition akin to Damasio—is extraordinarily malleable, and "tempera-ment is not destiny." In fact, he seems to be asserting that inner growth and transformation of personality are indeed possible.

In his groundbreaking book, Goleman refers to the need to be aware of one's feelings, to control anger, impulses, and anxiety. He also points out the need for empathy, proper expression of feelings, and constructive relationships. As well, he alludes to childhood as being very influential. In his sequel book, *Working with Emotional Intelligence* (1998), Goleman identifies a number of "emotional competencies"; on the personal level, he promotes recognizing one's emotions, assessing one's strengths and weaknesses truthfully, a sense of self-worth, self-control, honesty, responsibility, adaptability, motivation, commitment, initiative, optimism. On the social dimension, he identifies understanding others, assisting others to develop, persuasive ability, conflict management, change management, nurturing of relationships, and ability to cooperate and create group synergy.

With at least a general sense for the intended meaning of "emotional intelligence," it is well worth examining this phenomenon in light of the context of this book, and, particularly so, this chapter. When one is working, albeit perhaps not explicitly, with a one-tier model of the human (intellect, emotions, body and its sensations), an intelligence not of the intellect per se, and not specifically bodily kinesthetic, can be easily labeled "emotional." *There is no other component in this model to which to appeal.*

However, the term "emotional intelligence" may appear vague and confusing. One often hears warnings such as, "If you follow (solely) your feelings, you may ruin your life," and many are aware of the "fickle nature" of feelings. IS professionals in particular (where many are thinking and structured (judging) types) may be uncomfortable in knowing *how* to make their emotions "intelligent." Does this mean a rather superficial smiling and patting a colleague or a subordinate on the back instead of shouting angrily? Does this mean expounding optimistic phrases, albeit unrealistically, just not to sink morale? Is there a premeditated, willful modification of behavior required to act with "emotional intelligence"? Does one say, "I should act this way now," according to rules of EI, and choose to exhibit such behav-iors (this might be truly "acting")? Understandably, considerable confusion and ambiguity could abound in, say, making an IT department "emotionally intelligent" from a one-tier model of the human person.

However, if we *consider the two-tier model* introduced in this chapter, where the intellect, emotions, and body serve and receive the all-positive deep psychological core energy from the deepest inner self/being/core, *then the entire concept of "emotional intelligence" is immediately clarified.* The intelligence comes from the gratuitous innermost zone—the second tier. It then illuminates the intellect, the emotions, and the body, which, having access to this reservoir of wisdom, can thus display all the features that Goleman and others identify with "emotional intelligence." Both the mind *and* emotions become the servant of the inner being. Thus, in this view, the way to develop emotional intelligence in the adult is to work at removing (evacuating) the blockages between the intellect-emotions-body and the deepest inner self, and then to allow one's thoughts and feelings to be guided by the now-accessible inner energy where all the "awareness" should ultimately be found. In the child and adolescent, additional care can be taken to provide an environment so that the blockages are minimized (when the young person is adequately seen, heard, trusted, recognized, and safe to open, amidst vivifying relationships). With the two-tier model, the pure energy comes from deep within. Thus, *the reality of the deepest inner self/core/being/spirit is the actual root of one's emotional intelligence*, which, when appealing to psychology with the philosophic viewpoint in the perspective of Helminiak, may be better termed "spiritual intelligence" (although the term, if used without clarification, might give rise to a variety of misinterpretations).

The practical maxim, then, for the IS professional willing to develop in emotional intelligence would be: "unblock inwardly, connect deeply, be attentive to the deep self, and act accordingly"—simple, but not necessarily easy!

IT WORK FROM THE DEEP INNER SELF

From the first three chapters, the reader may have a reasonable idea how awareness of personality type and thinking, learning, and creativity styles may affect positively certain aspects of IT work. But *what is the major, practical contribution of rootedness in one's deepest inner self to IT effectiveness*, to the "bottom line"? Also, what evidence is there that deep-level consciousness is being considered within the IT area? This section will attempt to address both questions.

A Call for Transcending 'Rationality'

A major realization of the need for an infusion of inner human spirit, into techno-centered, computer-oriented society, was pronounced in the late 1970s by Joseph Weizenbaum, a prominent MIT computer scientist. This is the person who had earlier attempted to develop a computer-based psychoanalyst, but later admitted to significant misgivings about mechanizing an essentially human activity. In his book, *Computer Power and Human Reason* (1976), Weizenbaum exhorts:

"The computer—logical, linear, rule-governed, encourages a certain kind of thought process. Call it scientific rationality. We are now, as a society, close to the point of trusting only modern science to give us reliable knowledge of the world. I think this is terribly dangerous…the world is overly mechanical and…it invites the substitution of machines for human beings in fairly intimate areas of life…We should look upon warm, genuine human relationships…as endangered phenomena…We are deeply committed to a Faustian bargain that is killing us spiritually and may eventually kill us all physically."

He then offers the seeds for a new vision:

"What it takes, for an authentic transformation of our society is an energetic program of technological detoxification."

Weizenbaum was indeed met with criticism, skepticism, and cynicism within and outside of his field. However, due to his stature and prominence, his remarks did not go unnoticed. Do Weizenbaum's remarks, even in part, have application to the IS profession today? Many would say "yes." In the IS profession itself, many are remarking about information overload, "analysis paralysis," impending burnout, and need for work-life balance. To quote psychologist Mary Riley, "High touch has not kept up with high tech." However, how can a person or even the IS profession in general respond to Weizenbaum's research? I recall speaking with a senior IT consultant from Germany who had spent a considerable part of his career with a prominent international IT organization. He saw in Weizenbaum an *anti-intellectualism*. "What does he mean, we are thinking too much?" Indeed, with a one-tier model of the person, which is very predominant at this point in human evolution, particularly in industrialized societies, what does one have when he moves from the intellect? Unchartered and unpredictable emotion, sensual gratification? From where can one grow?

However, with the two-tier model, which admits to the inner core at its own, deepest level, Weizenbaum's insights can then be viewed as *trans-intellectual*. We descend from the intellect to the deepest self, connect our intellect to it, and have the intellect live at the service of this deep core energy. The German IT consultant, however, had some questions here, too. He cited many examples of people who tried to analyze "what lies within" and found that it was mostly "not good." Again, two-tier model to the rescue! It is postulated that what was "not good" on the descent from the intellect were the *wedges* in the model, the inner blockages from awareness of the pure and all-good energy within. Against a background of the two-tier vision, Weizenbaum is attacking not "too much thinking," but "too much disconnected thinking," i.e., knowledge without wisdom. In this light, he can indeed be a prophet for IT today.

Psychological Robustness and Interdependence

The main benefits to IT of growth into and work from the deepest inner self is *psychological robustness*, capacity for psychological resiliency in the face of overwhelming challenges. Ordinarily, when reacting to stressful, unfamiliar circumstances, unfavorable emotions such as anger, frustration, or fear of failure can take over the psyche and largely disempower the intellect. We can imagine, at this point, a "Psyche-refresher" machine. One comes up to it, attaches a large vacuum cleaner-sized hose (coming from a large canister with a foot pedal) to one's chest near the diaphragm, and steps on the pedal. This is the *hose of psychological energy*. All of a sudden, deep within, the exasperated IT worker is being filled with profound peace, harmony, self-acceptance, self-confidence, motivation, encouragement, patience, tolerance, and creative powers. Now, all these psychological realities are available to his intellect (as well as his feelings and body).

Indeed, the "magic canister" with the hose *does exist*. It is present deep within, at the level more foundational than the intellect, emotions, or body. It indeed provides the empowering psychological energies that accrue when one has grown sufficiently, to be able to "drop oneself into" one's deepest core energy in the face of external storms in the course of IT work. Availability of such an empowering inner resource depends on maturity in growth. This affects significantly one's essential, emotional orientation to the required efforts of one's daily work

A person can relate emotionally to one's work involvements in three possible modes:

1. *Dependence:* One is *close* to his involvement, but *not emotionally free* of it.
2. *Independence:* One is *emotionally free* of his involvement, but *not close* to it.

3.   *Interdependence:* One is *very emotionally close* to his involvement, but simultaneously capable of being *emotionally free* (detaching) from it.

IS work requires precision, creativity, and solid commitment. Most IS managers would indeed like to see their subordinates intensely involved in their work. However, why can such involvement in the face of rapid change, time pressures, economic/organizational instability, etc., lead to burnout? Is the answer to be "free and not close"? Again, it depends largely on *which model of the person the IS worker has internalized.*

From a one-tier perspective, it is possible to live in dependence and in independence from one's work, but likely it requires two-tier consciousness to live in interdependence, being both close and free. All depends on one's perceived main identity. If one's work and its successful results are the main source of one's identity, security, and meaning, he can be very close to his work, while fearing failure and blocking his inner life force during challenging times. On the other hand, if one's family, hobbies, wealth, etc., form his identity, he may treat his work as simply "a job." When the work environment exerts considerable pressure, an emotionally independent worker may quickly run out of the extra motivation and energy needed to stand up to the task. When job stress induces inner blockage, he may have it relieved, from day to day, from his out-of-work involvements, but his capacity to draw on empowering psychological energy from within is nearly nonexistent; moreover, an independent worker does not have the commitment and resulting psychological drive to excel. He has largely "turned off" reacting to external pressure to perform, but he has not "turned on" his connection to the power within.

"I Am - Therefore I Think"

Having adopted, both conceptually and experientially, the two-tier model for life and work, *the interdependent worker can be the IT Department's greatest human resource.* This type of IT employee derives his fundamental energy and identity from *being.* Connecting to the source of pure fulfillment, he is, in essence, fulfilled. He then works *from* fulfillment, not *for* fulfillment. He works *because* he is energized, not *in order to be* energized. He realizes that what he does (or who his family is, or the size of his bank account, etc.) is not who he is—he *is* because he is connected enough to his deep inner "*is*ness" (where there is nothing negative and nothing missing for essential joy—i.e., life energy). To the question, "Who am I?", he can answer with rooted certitude, "I am." His work then becomes an *expression* (not simply in principle but in reality) of who he is. He is living "I am—therefore I think," and such connected thinking will be crisp, effective, creative, and flexible.

Thus, in identifying the three types of workers, an IT manager may begin to wonder, while GUI design, code testing, or data modeling are going on among her subordinates, *what is likely happening for each of them at an inner level*? What are they feeling, how are they reacting emotionally to the tasks and challenges at hand? More importantly, she may wish to assess which persons are capable of being very involved, yet not attached in dependency (not addicted to their work), in short, in PRH terms, which persons are indeed "living their talents" in the workplace. This is emotional intelligence in action!

IT Worker Identity

Having recognized the ability to draw on a deep inner resource to give buoyancy to the intellect and the feelings, it is now in order to identify and examine *specific features of IT work that the inner core energy can affect* significantly.

Firstly, living and working out of a two-tier consciousness is bound to affect significantly an IT worker's *identity* and *self-concept*. Considerable literature has expounded on the limited capacity and desire on the part of the systems "techies" to understand and address adequately, through effective communication, the business needs of the users. Often, the "techie shell" is a source of deep emotional security for an analyst or programmer; any personal or professional insecurity can be further bolstered by the detached genius-expert image. This is the function of one's *ego*.

When information systems were more basic in both their content and organizational impact, such an image of exclusivity may not have had as much negative impact. Today, with e-commerce, multimedia, and systems for competitive advantage, considerable teamwork and back-and-forth *effective* communication with business management is mandatory for systems developers.

An "operational connection" to one's inner core energy will *transform one's ego*, allowing it to tap into the deep, pure, vitalizing energy of the inner being. One will then, automatically and without conscious effort, be able to experience oneself as respectful of others, open to others, caring for others, interested in others, and delighting in serving others. While one will have a profound satisfaction in a truthful assessment of one's developed IT skills, the inner core also provides gratuitously an attitude of "felt humility," that is, acknowledgement of a truthful perspective of both the value and the significant limitation of one's skills and abilities. Accepting realistically and acknowledging one's limitations is not at all intimidating to an inwardly connected person. One is always humanly limited, *but never deficient*. On the level of the deepest inner core is everything one needs for a sense of worth, dignity, empowerment, genuine confidence, serenity, and optimism. Such essential traits do not "run dry"; on the contrary they are always capable of expansion within our consciousness, to the degree that we are able to allow them in.

Attaining "Soft Skills" - With Little Conscious Effort

With an inner growth approach, one does not have to modify behavior consciously. Modification comes from a one-tier consciousness ("I better start to listen to that Marketing VP if I want to keep this job"). With inner growth, the willpower is used only to choose to unblock inwardly in truth during the growth process—the deep life energy, heretofore unknown and unexperienced, now fills one's consciousness gratuitously, inducing one to change one's behavior automatically, genuinely, and almost effortlessly. *Grasping, experientially, this truth can make a profound impact on the professional and personal life of an IS professional.*

Thus, growth-in-being can significantly transform an "IS techie's" identity. With a new consciousness in IT workers, *teamwork and user communication* will unquestionably become more effective, by a "quantum leap." With a one-tier consciousness, much communication is hampered by emotional "self-protection"—one does not wish to expose one's ignorance and limitation in understanding, one does not want to show one's professional and/or personal insecurity, one may be threatened by approaches or styles different from one's own. With a growth-induced two-tier awareness, such defense mechanisms are unnecessary and would not even be considered. With such an awareness, one knows that one can dissipate self-consciousness with *consciousness* of the pure, all-encompassing energy emanating freely from the deepest inner part. With an awareness of the deepest self and connection to it, it is very easy to admit, fully and truthfully, as appropriate, one's shortcomings and limitations, because one is simultaneously aware of one's indisputable value, goodness, and worthiness which come from deep within.

With an expanded consciousness and self-concept and, thus, a transformed identity (from techie to a multidimensional *information resource facilitator*), the new IT worker can indeed be effective in the current system development climate. Through his/her now natural capacities for cooperation, respect, listening, flexibility of viewpoints, fairness, optimism, and emotional resilience (all aspects of two-tier consciousness), coupled with an awareness of significant one-tier psychological factors as outlined in the first three chapters, such an IT worker would become truly emotionally intelligent.

In *teamwork*, he would gladly listen to and assess opposing viewpoints, and would be secure enough at the deepest level to welcome constructive criticism. Out of a profound sense of self-respect, the sense that comes freely from deep within, a developed IT worker could indeed make users feel valued, and have their needs and reservations respected and addressed adequately. A project manager with considerable growth in the inner self could be sensitive to the unique talents and psychological needs of subordinates. She could truthfully express firmness or disapproval of a person's behavior without anger or fear and without having the person feel undesirable. Through her openness and genuine humanity, she could invite openness among subordinates, generating considerable *trust*.

If a manager addresses a subordinate from the intellect (alone), the subordinate's reaction will also likely come from the intellect. If the manager communicates with emotion (perhaps superficial, e.g., "that's the stuff"), the subordinate will respond likewise. However, if the manager connects from her own deepest inner, authentic self, this will prompt a response from the same deep level, especially if the subordinates have been undergoing their own process of inner unblocking and deeper connecting. Without doubt, truly "being-centered IS teams" of the future will indeed be a sight to behold and a model for work in the post-modern era.

Conflicts with a team or with users can be much easier to resolve from an inner-center consciousness. There will never be attacks on a person or his dignity out of a sense of bruised ego, since inner-centered persons do not rely on ego for security and strength. The deep, freely available inner strength will allow for considerable flexibility in generating new approaches, which parties in conflict will be able to accept—people will not need to "defend their turf" so rigidly as if their life, self-worth, and meaning depended on it.

In the course of IT work, frustration, anger, and disillusionment, which are humanly natural, will not disappear from people who are considerably integrated. However, the negative effect of these emotions on the thinking capacity, motivation, and physical health will be considerably lessened, since these emotions will not need to be inappropriately expressed or unhealthily repressed, but will be offset and largely "melted" by access to a more centered, profound, powerful, peaceful, and energizing inner force.

IS Work and Control

The issue of *control* is often central for many IS workers. With many system developers having come from a third-generation language programming background, structure was paramount and control over one's program was very possible. Now with realities such as unrealistic deadlines, ongoing changes in requirements and development approaches and frequent organizational restructuring, a sense of control over one's work tasks, one's competency level, and one's career path is often eroded. For many structured personalities who may have relied on such control for most of their essential personal security, such an erosion can indeed be traumatic.

With an active connection to one's deep inner self, which feeds one's consciousness with all the deep essential security one needs or wants, one can be emotionally *interdependent* with work structure and control rather than being emotionally *dependent* on it. One can be close to structure, work with it, maintain it, yet be simultaneously emotionally free of it. Then, a disruption in structure and loss of control, while it may generate discomfort at one level, will never disrupt one's inner sense of deep security, one's conviction of being "OK." Thus considerably less energy is being drained, leaving the energy for productive responses to the IS work reality.

While considering control issues, the topic of *locus of control* has been mentioned in management and MIS literature. This can refer to whether, in a given work situation, one feels that control is outside of himself or within himself. An external locus of control can generate experiences of being a victim of circumstance. An internal locus of control enables more options. However, until now, internal locus of control has (presumably) meant, on a one-tier level, control by one's intellect and will. Now, if one is working from a two-tier, inner-core consciousness, the *concept of locus of control can be extended* for both research and practical considerations. If, indeed, in a given work situation, the locus of control is internal, *where* is it internally? Is it in one's intellect, emotions, or the deepest inner core?

Dealing with Ongoing Change

Related to loss of control are the needs within IS for coping with *ongoing change* and the need for constant, substantial learning. We have seen that one can be emotionally *close but not free* or one can be *close and free* to one's work efforts. In dealing with change and the need to unlearn and relearn, the IS worker does need to be closely involved, yet cannot afford to become too closely attached to a particular paradigm, methodology, or acquired skill. Considerable distress in change results from needing to let go emotionally of what has become comfortable, fulfilling, and even affirming. However, the degree of distress in change will depend on the intensity of the initial attachment.

To illustrate, one can enjoy fully the use of a cottage at a Florida beach resort for one month. At the end, there is not much emotional distress from needing to leave the cottage, since it was never emotionally accepted as a permanent home and since one knows one has a permanent home to which to return. Similarly, particular system development software can be learned and enjoyed for the period it is used; when it is changed, one knows that many essential skills developed will be transferable, but much more importantly, one realizes, through a deep inner connection, that one's "true home" is not being taken away.

Work/Life Balance

Another area of concern to IT workers is *work/life balance* and the prevention of burnout. In the area of Internet and competitive advantage, the demand for an IT worker's time and energy seems continuous. Wisely, the profession has recognized that life balancing is beneficial and necessary. Yet, for true regeneration of mental, emotional, and physical energy, the work/life balance need not involve only horizontal shifts from one intellectual activity to another, or to emotional or physical pursuits. A horizontal, diverting shift in attention and energy may be quite adequate for one who is simply "tired." However, if the work environment has occasioned numerous negative emotions and may have also challenged significantly one's inner security and perceived essential identity, such a horizontal shift may not be adequate.

A middle-aged 3GL programmer working with stringent deadlines, who also fears skill obsolescence, is resentful of his manager's condescending attitudes and feels that his life has become largely without meaning. To "recharge," this person may not benefit significantly and deeply by taking up a new sport or ensuring that he has one evening a week to "drink with the boys." His work environment is eroding his sense of basic worth and security. His work/life balance must, most of all, be an *intrapersonal* balance, a vertical balance, and it must allow for time and inner space to access the deepest core energy, where there is possibility for genuine, profound revitalization and inner reorientation.

An article on maintaining work/life balance in *Information Systems Management* (Spring 1996) suggests that, for lessening the threat of burnout, one should: i) develop a strong set of core values; ii) create and maintain an energizing, personal vision; and iii) place one's work within the larger context of one's personal vision. With practical adoption of the two-tier awareness, the IT worker can go beyond developing a *vision*—one can develop a connection to a deeply energizing inner *consciousness.*

Rootedness in one's deep authentic self provides positive energy to the emotions and intellect. One then can, by his authentic presence, *motivate* others genuinely. One can also be inwardly free enough to attempt new, creative approaches to established work tasks. As well, deeper inner consciousness has positive effects on the *planning of one's IT career.* Following one's deep authentic self, one is motivated basically by truth (as seen by the inner eyes) and not by image or external goals. Such an IT worker will seek to know, truthfully, his natural talents and shortcomings; he will be content to work at his current level until genuinely motivated, in all parts of him, to seek another level. He will avoid career management through restlessness, "drivenness" to constant progress, or addiction to "mental highs" in order to drown out a basic inner malaise.

A rooted IT worker living mostly with her intellect, emotions, and body at the service of the deepest inner-core energy will become indeed emotionally intelligent at work. She will have the courage and ability to access and understand her feelings, and to intuit about the feelings and attitudes of others. Moreover, the inner wisdom of the being will serve as a valuable guide in specific situations. She will "just know" what words to use to diffuse a client's anger, she will "know" how to handle an apprehensive user, and she will "know" how to correct a subordinate without bruising his ego. Considerable IS literature calls for development of "soft skills." Such a development is best addressed by comprehensive integration, by connecting an informed intellect, aware feelings, and a healthy body to the deep core energy—by promoting "personal wholeness" within IS work.

At this point, it is worth noting that in this chapter it is maintained that for a full, integrated, optimal human growth approach to enhancing the work of the IS professional, *growth in the deepest inner self (essence, spirit) is indispensable and likely most significant.* However, a person with inner stability, wisdom, and balance can still benefit significantly from valuable insights at the intellectual or emotional (first-tier) levels. This is the level that most of IS soft skill development literature addresses today. Thus, to become more effective in IS, one can become more intellectually knowledgeable, more emotionally aware, but also more spiritually connected. All components have their place. It is a challenge to future research (perhaps on "Psychoinformatics") to define more specific relationships and dynamisms between the components of the human quadrinity in the course of IS work.

Critical IS Competencies

In the Spring 1996 issue of *Information Systems Management*, authors Longenecker, Simonetti, and Mulias identify, based on a study of 75 experienced IS professionals, 10 key survival skills for IS professionals:

1. Ability to balance technical and non-technical skills
2. Strong interpersonal and communication skills
3. An orientation towards business solutions
4. Ability to be an effective team member
5. Strong project management skills
6. Effective planning and organizational skills
7. Strong analytic and creative skills
8. Flexibility and adaptability to change
9. Responsiveness and a customer orientation
10. Ability to function as a teacher and a coach

As we examine these critical success competencies, it is well worth, at this point, to consider where a specific inner shift from living a mind-emotion-body awareness to living from a mind-emotion-body-inner-core/self/spirit awareness (the content of this chapter) would have significant influence.

As for #1, deeper awareness allows one the inner freedom and confidence to move beyond clinging to "techie" exclusivity for security and identity. The inner process gives rise to a willingness to assess oneself truthfully and to be open to multidimensional growth, including personal and business orientation (#3).

Since interpersonal communication requires a deeply rooted confidence, as well as a respect for and interest in others, and an ability to recognize others' feelings, such resources are available upon contact with the inner core (#2).

Teamwork (#4) requires compromise, respect, genuine humility, and enthusiasm; all are made possible by tapping in to the central talents "to be."

In project management (#5), a core-connection may well help in motivating personnel, releasing their potential energy, resolving conflicts, and supplying courage to be firm, yet positive, as well as in having the strength to make difficult decisions.

Planning (#6) is most effective when it is based on truth, rather than illusion or unrealistic optimism. Planning requires wisdom in addition to technique, and the deep self enables this capacity.

Creativity (#7) springs from inspiration and inner courage to risk. It is rooted in one's deep spirit.

Flexibility and change management (#8) are major areas where a deeper psychological robustness makes a significant difference. The deep inner consciousness that, at the root, "all is OK," provides such groundedness and confidence to actually embrace change as welcome, rather than to resist it apprehensively.

A customer responsiveness (#9) requires humility and the inner freedom to be open to hearing and acting on the needs of others. Inner freedom is possible when one does not cling internally for security, but rather allows the deeply rooted energy of security to surface to him.

An integrated person is by nature genuine, fulfilled from within, and motivated to communicate this fulfillment by assisting others. She is capable of self-transcendence and does this very naturally; this indeed can make an excellent mentor (#10).

Would not an IS worker developing such "survival skills" want to make use of *all* his/her available resources? Re-examining, at this point, earlier quotations from Siegel, Borysenko, Dyer, Carpenter, and Zukav and Francis may indeed lead one to consider *the tangible potential and real IS empowerment that can arise from the level of the human spirit/inner core*. Ultimately, however, such an acknowledgement can only be made after an inner, personal experience—no one can taste an orange for you. Thus, when applying growth-in-being specifically to business firms, PRH has stated: "Firstly, though, the chief executives and persons in management must choose this integral growth for themselves and experience its benefits."

References to the Inner Self and EI in IS

As we observe the increasing number of authors and personal growth programs addressing and promoting the "deepest inner power" (under various labels), we realize the initiation of a significant shift in consciousness is an increasing part of the post-modern world. Has such a shift already affected the IS field specifically? We now look at specific, noteworthy incidents.

In 1988, U.S. management scientist and academic Robert Thierauf, in his book, *New Directions in MIS Management,* in a chapter on "Motivating MIS Personnel," describes self-actualization as "realizing one's ambition," that is, self-fulfillment. He then questions whether "I do my thing" is the final goal, the farthest one can go in personal/professional development. He answers, "No," and proposes further stages of *mutual actualization,* where "two or more individuals enrich one another and in doing so, they grow both as individuals and as a group," and then *self-donation*, when one has a attained a level of expertise and satisfaction, and now offers assistance to a junior person in her growth. Thierauf also declares that "change and growth are not necessarily pleasant processes. Sometimes there is sadness, sometimes 'raw' pain. But what happens to us is real. It deserves our full attention." This may well be seen as a precursor to the growth-in-being promoted in this book.

Several IT industry newspapers have carried relevant columns. In 1996, *Computerworld* called for "emotional literacy" among IT personnel. Articles on "emotional intelligence" have appeared in *Computing Canada*. In one such article, a president of a Toronto firm specializing in increasing human performance says, "Promoting emotional intelligence can be the key element in assisting IT employees in developing interpersonal skills." He believes that it is the emotional relationships and connections workers have with their peers and supervisors that make the difference in their productivity and loyalty to the company. In another article, entitled "Info Tech Workers Emotionally Weak," a psychologist defines EI as a "measurement of how people cope with life." He administered an EQ (emotional quotient) test to a number of different professionals and found IT workers had a slightly lower score than most other groups. He had used the Bar-On EQ-i test, developed by Israeli psychologist Dr. Reuven Bar-On, to gauge the coping abilities of 150 IT professionals. It would be interesting to apply such a test one time, and then again five years later, to a test group of IT workers who would seriously undertake a growth program such as PRH for at least four years.

In a recent issue of the *Cutter IT Management Journal,* in an article entitled, "XP and Emotional Intelligence," author Kay Pentecost relates an incident where her development team, wishing to please its manager, took on a dubious goal of trying to improve a working demo for an executive in a very short period of time, and failed to have even the basic demo working on time. In hindsight, she acknowledged that the desire to please their manager clouded the team's judgment as to the feasibility of his request. She pointed out that, "In this group of supposedly logical developers was a lot of negative emotion: blame, fear, anger, frustration, and some amount of despair." Then, some time later, a similar situation arose, with an inappropriate request for an "express enhancement" to a working demo. The author, tapping into her inner strength and wisdom, truthfully told the supervisor that the team could not do it, in spite of angry reactions. The original demo was presented to the VP, without desired enhancements—the VP was impressed.

Upon reflecting on this scenario, we firstly notice "emotional intelligence" as a topic in an IT management journal. Secondly, with the exposure from this chapter, we can appreciate that it took emotional awareness and the connection to the deep inner truth (which is present in our core) to pronounce a correct decision in the face of opposition and pressure.

Another approach, observed in some software development environments, is the availability of *meditation rooms*. We can recall remarks from Dr. Bernie Siegel: "I first got to know this perfect core self through self meditation…however you get there, you'll know you've arrived—it's like coming home." Thus, a main purpose of meditation (e.g., repeating a "mantra") is to quiet the "noise" of the intellect to allow space for the inner core energy to surface to consciousness.

Indeed, in the article, "The Use of Meditation-Relaxation Techniques for the Management of Stress in a Working Population" (*Journal of Occupational Medicine,*1980), researchers Carrington et al. report, where telephone company employees practiced meditation twice daily for six months, there was an effective reduction in symptoms of stress, such as depression, somatization, and anxiety. Also, from participants' spontaneous comments, the most frequent benefit was "improvement in cognitive functioning." Other comments have included: "My reasoning process is clearer"; "I think, remember, and organize better"; "…increased interpersonal awareness"; and "…do not feel so defensive in my relationships with other people." These effects may indeed be interpreted as resulting from having the mind at the service of the deep inner-core self.

However, it is also worth considering an earlier-mentioned caution from computer scientist Orest Bedrij, who advises that meditation not be practiced apart from: i) releasing of significant emotional blockages and inauthentic values, ii) "purity of heart," iii) an attitude of in-depth love, and iv) lack of excessive attachment (emotional "clinging") to specific factors in one's life for emotional security. Thus, meditation would likely have varying degrees of effect on the work of software developers. Considerable research in this area may be of interest and of significant assistance.

Two specific, individual comments from an IT context are also worth noting, since they may be considered as "pioneering." In a 1998 weekend edition of the *Boston Globe,* an advertisement for IT jobs in an insurance company contains the following:

"Sometimes you have to listen carefully, to hear the loudest voice. It's the voice inside. Stop ignoring that voice—it's coming from your soul."

Such a direct reference to the deepest inner core can now be more clearly appreciated in light of Zukav and Francis' contention that authentic power comes from "alignment of your personality with your soul." Such an orientation can induce human resource recruiters and IS managers to try to assess not only whether a candidate/employee is highly motivated, but whether such motivation is largely resulting from an insatiable, addictive drivenness of unbridled ambition or whether the motivation comes from the deepest authentic self. It may be a challenge for researchers to develop and validate a questionnaire that may try to assess such an important factor.

The other pioneering comment comes at the start of the popular IS text, *Introduction to Information Systems—Essentials for the Internetworked Enterprise* (1998). Author James O'Brien tells his readers, *"May you love the Light within you, And in everyone you meet, And in everything you experience."* A clear reference to the deepest inner self in such a context is indeed enlightening. It can certainly be understood as a call to a deeper inner awareness while developing an IS orientation. Might it happen in the near future that self-awareness and inner-core-led empowerment, within appropriate case studies, may become part of IS management curricula?

A 1998 article in *Time* magazine, entitled "Get Thee To A Monastery," identifies a developing trend for overworked professionals to retire to rural monasteries for relaxation and inner refocusing. There, nature, simplicity, pervasive silence, as well as the attentive presence and rhythmic chants of monks disciplined in living a deeply rooted and authentic life, offer "the most refreshing vacation going." The article describes a 45-year-old computer specialist who considers herself "spiritual, not religious," who has decided to come back to an upstate New York monastery each year for her only vacation away from her job. Then, she finds inner peace and can "sit, think, and pray." It is also interesting to note that in the book, *Using SAP R/3* (Que, 1997), in the chapter on "The Human Side of SAP Implementation," meditation and religion are identified explicitly as avenues for managing stress.

In 1994, a computer consultant in Louisville, Kentucky, published more than 40 articles on "The Human System" in a computer newspaper. He was also involved in giving "Human System" seminars to help people to live more satisfying personal and work lives. In one article, he points out how "our spirit controls our mind, and, in turn, our mind controls our body," and encourages readers to "control your mind with your spirit {deep inner core energy}." This same person has been promoting the slogan, *"We are spirits having a human experience,"* as opposed to humans having a spiritual experience (he had even been selling t-shirts with such a motto at an IT convention).

He provides feedback received from his listeners, e.g., "It gave me a whole new understanding of myself."

While compiling incidents of specific reference to the deepest inner self in an IT context, we can revisit briefly the case of the keynote speaker at the 1995 National Conference of the Canadian Information Processing Society, who urged IS professionals to relate to others at work not in a mode of confrontation, but in a mode of love. Considering this beside the PRH distinction between popular, sentimental love and in-depth authentic love (made possible by connecting to the inner core energy), one can understand the consultant's advice as a call to a shift in consciousness from a one-tier to a two-tier model.

At that same CIPS Conference, I was offering a 30-minute presentation on "Personal Wholeness for the IS Professional." I had defined "personal wholeness" as a "connectedness of the mind, body, and feeling to the inner being, such that the energy of the being can express itself through the other three parts." I had been told to expect about 20 attendees, but was fortunate to have been allocated a slightly larger room—the actual attendance was 75. As well, in my "Wellness for IS Workers" seminars, out of four modules on the mind, the body, the feelings, and the inner core, the material in the last module was consistently evaluated most highly of the four (perhaps such a seminar appealed most to intuitive feelers at the start, since their main need is to be authentic to the deepest self). With arguments for relevance of two-tier consciousness to IT work and evidence of particular incidents of involvement, it may be of use, now, to consider specifically what an individual and his/her employing organization can do in this regard.

FOR THE BEGINNER

While it is, for most, likely easier to accept in principle the existence and benefit of personality typing systems or thinking/learning style classifications, the idea of a deepest core, as an entity separate from intellect, feelings, or body, is likely to appear much more novel. Consequently, after having read the introduction to the deepest self presented in this chapter, a beginner in this area might want to read and think more. The books referenced earlier, notably those by Hoffman, Dyer, Borysenko, Siegel, Helminiak, Zukav and Francis, McGraw, and the PRH Foundation, as well as EI books by Goleman, could provide the intellectual understanding to form an adequate conceptual foundation.

From then on, how one proceeds may depend to some degree on one's personality as well as one's perceived need for this level of inner growth. It is likely that for many (some NF types likely excepted), this will not be the *first* dimension of psychological growth. Initially, they will likely be more comfortable developing an awareness of their personality and thinking style. Whenever and however one arrives at the point of seriously considering growth in the deep self, the following pointers may prove useful.

It is suggested that you reread the main parts on the inner self that have been presented in this chapter. Please remember that you cannot will or think yourself into your inner core. Review your personal history and non-judgingly list possible areas of inner blockage. As well, discover truthfully specific incidents, isolated or recurring, where you may have indeed experienced an opening to your innermost zone. With such awareness, plan, in general, your growth approach. Will you have a growth facilitator, formally or informally? What supporting aids will you employ (e.g., meditation, nature, music, etc.)? Will you have a "partner-in-growth" doing similar work alongside you? If it is possible, in your area, to attend a weekend workshop on opening to the deep self (e.g., the PRH workshop, "Who Am I"?), it would be wise to consider.

Bear in mind that growth *will* involve deep feeling and that it will not proceed linearly. One day it may appear that you have made a major break-through in inner awareness, and the next day most of it may seem eclipsed. Patience and perseverance are definitely required! Apart from temptation to grow "by the sweat of one's brow" using mainly willpower and analysis, there are some other "temptations" that may divert a novice from the path of genuine inner discovery. For a "thinker" personality, it may be easy to confuse emotional excitement about having discovered, intellectually, a new model of human functioning, with genuine opening to the inner core. The excitement will be more on the surface and can recur often in initial stages. An experiential breakthrough to the second-tier will at first occur rarely, in glimpses. However, such an experience will indeed be powerful (you'll *know* you have shifted levels). As one IS professional remarked:

"It is like connecting a large vacuum hose just under my breastbone; the hose will then pump in, totally gratuitously, the most intensely profound feeling of fulfillment, confidence, self-acceptance, optimism, and peace—most of all, you realize this is all being given freely and is not tied to anything that is happening around you."

He had indeed had a breakthrough experience!

For the "feeler" personality with naturally warm feelings, there may be an inclination to confuse experiencing such feelings and perhaps projecting them onto others as an experience of the deepest inner core. The deep self can express/vibrate itself through feelings, but feelings (like thoughts) can also operate on their own without substantial deeper-level connection. At times, staying in warm feelings may prevent one from willingly expressing deeper, more difficult feelings, which are necessary to break through.

Please, also, realize that it takes courage to evacuate emotional blockages rather than repress them and then compensate for this blockage to fullness of life. Take heart in your choice to evolve! Most importantly, one should realize that, in general, our society has not evolved to seeking this deep level of living. To quote PRH: "Our Humanity is extremely underdeveloped in the area of in-depth riches." Parents, teachers, mentors, etc. have not themselves received enough inner awareness so as to know how to facilitate this in a growing person. Thus, our inner "blockages" may have resulted even from a "normal" development—normal, however, for this age and not in the absolute sense. Yet most people have not had sufficient attention focused on all aspects of their person that were waiting to unfold. Children may not have been deeply, emotionally accompanied though various essential dimensions of their adolescent development. Schools may not have supported enough of one's creativity or unique thinking style, resulting in inner blockages of intended awareness. Families may have unreservedly imposed cultural norms, in a "dictatorial" fashion, stifling one's need to discover, discern, and to decide for himself on key aspects of his identity. Thus, we may have blockages from our deepest self because we are humans living at this point in humanity's inner evolution and not necessarily because of an unusually traumatic past. A welcome motto here, for this type of inner growth, is that "truth liberates," and it also empowers. Encouragingly, there are some signs that humanity in general may be at an "inflection point" in its inner awareness. To quote, again, Zukav and Francis:

*"The old species explored the physical world and it created security by manipulating and controlling what is discovered... **The new species creates security by looking inward to find the causes of insecurity** and healing {unblocking}them. This is the path to authentic power" [bold added].*

As you notice even small changes in awareness, it might well be worth writing down what you are experiencing. At the moment of a more significant inner breakthrough, consider and note how, *at that moment*, you see what have been your major current difficulties, particularly at work (e.g., an analyst has had considerable recurring conflicts with a domineering user and has harbored considerable anger and fear towards him—at the moment of inner connection, though, this newly available *free energy* gives him the potential to see this user as frustrated and insecure, and to feel, within himself, the power to work with this person in a "close, yet free" emotional mode (here is the root of emotional intelligence in action). Initially, though, the moment of connection may not stay with you. But it will recur. It may take only one such significant emotional shift to genuine empowerment to convince a beginner, with unwavering certitude, that this type of growth is indeed "for real." Then, it would be wise to connect with others, at work, who have been similarly "breaking through" and to share, particularly the effect of such awareness and inner power on specific aspects of one's work in IS. With enough people choosing to discover and apply daily the empowerment made available from the deepest inner self, physical and virtual groups, newsletters, and websites may soon emerge.

Organizational Involvement

What should be the role of an IT department regarding growth in the deepest inner self? This will depend, of course, largely on its management's views on the usefulness of psychological awareness in IT.

It is unlikely that IT management's promotion of soft skill development would start with the deepest inner self. However, after considerable acceptance of personality awareness and recognition of tangible evidence of its impact on IT work, some managers may indeed take note of the potential of deeper inner empowerment. In all likelihood, such managers would be NF types, whose deepest personal need is for profound authenticity. It would then be wise for such managers to enroll in inner growth programs themselves and begin to notice significant benefits. A network of such pioneering managers (perhaps using a slogan, "IT for IT—Inner Transformation for Information Technology") can assist their fledgling vision.

After perhaps two or three years, such individuals in middle-to-higher management (who may indeed have been initiated to such a perspective by this chapter) may give presentations to their staff on how such growth has helped *them* at work. Employees may then be invited to explore such potential empowerment, and some organizations may even offer subsidies to people willing to participate in growth programs. It will be very important to continue documenting specific benefits to IT work and disseminating this information broadly. One will not be able to (and should not) "legislate" growth-in-being, but in time, it is hoped that results will speak for themselves.

FINAL CLARIFICATIONS

When I have presented material in this chapter, inevitably people would listen. They may have questioned what they heard, they may have had difficulties relating, or they may have been indeed glad that finally this topic is being brought out. Because the notion of a "deepest inner psychological reality" is new to many, I would receive various questions after my presentations. Following are a few examples of such questions followed by attempts at answers.

Is awareness of a totally positive inner zone really a good thing for IT work? Wouldn't "connected" people just be complacent? Don't people create because of a drive for fulfillment?

The goal of an integrated life is to be true to all dimensions of oneself. The deep inner self provides essential fulfillment. However, this is a live energy which yearns to express itself through one's peripheral talents. If a person is a gifted programmer, he will be truthfully motivated (not addicted to, not "driven") using his ability. A person working *for* fulfillment is "close, but not free" with his work—he is much more affected by setbacks and conflicts; a person working *from* fulfillment is able to be "close and yet free" with his work—he is much more focused and resilient after setbacks, since setbacks affect his periphery, but not his central identity or existence.

I can't believe anything I can't prove scientifically. Where is there proof of an "inner self"?

This leads to much broader philosophic issues such as the definition of "science" itself and whether it can indeed be truly "unbiased." Also, is all of human existence part of science, or is science part of a larger sphere of human existence? Note Weizenbaum's misgivings about the universality of "scientific rationality."

Also, it is quite accepted within the academic field of MIS, for example, to study the effects of "trust" on the MIS-user relationship. Is "trust" scientific? How do we prove that one is, indeed, trusting?

The discipline of psychology has a positivist viewpoint and a philosophic viewpoint. You may wish to investigate how they differ. As well, consider the mounting empirical evidence regarding "empowerment" which is being written about in the cited literature. In the end, though, it remains mostly a matter of personal experience. One person cannot taste an orange for another.

Why don't we just face up to the fact that we are human and have limitations? This appears to be a New Age attempt promoting exaggerated human powers, at worshiping the human much beyond his real capacity.

Inner growth aims at discovering the truth and dignity, all that is fully human, in intellect, body, emotion, and spirit (deep self). The reality of the deepest core has been part of religious and cultural wisdom from early times onward. However, in the Western world, particularly after the Industrial Revolution, the intellect became dominant at the expense of considerable awareness of the deepest inner core. As long as the intellect was able to control, people did not notice what had been eclipsed from them. Control provided much inner security. Now, in postmodern times, for many people control has been significantly eroded. But, it is possible to rediscover and open to the deepest inner (spiritual) energy that provides more solid and satisfying security than a sense of control. It gives a new type of power, while offering freedom simultaneously.

In this view, the deepest inner core is not limited—it is capable of infinite expansion, but our humanity is limited in receiving and using it. A true contact with the inner self provides peace, confidence, yet profound humility. This philosophy has been present in many sources of human wisdom, and some of it may also be present in the eclectic "New Age" movement. However, that is not the central issue.

I was raised in the typical North American culture—porridge, bacon and eggs, apple pie, baseball, etc. I like myself that way—I cannot see myself growing long hair, wearing sandals, smelling incense, and being spaced out "in contemplation" in order to be a better IT worker.

The goal of growth is to discover the best self that *you* can be, from the inside out, and to "put it all together." It is to be the most authentic human person you can be and to bring this person in a "close but free" relationship with IT work. Human authenticity is about truth, joy, and genuineness, and not about clothes or, necessarily, food. While meditation may indeed be a way to clear the intellect so as to allow the deeply energizing life force to begin to surface, it should give one's personality more solidity, flexibility, openness, joy, and naturalness. A truly inwardly connected person is deeply, inwardly attractive, but *never* "spaced out." It can be, however, that one is "meditating" without inner unblocking. Orest Bedrij, the computer science researcher who fasted, prayed, and meditated before discovering the equations for a type of parallel processing, warns against "stillness without a pure heart." This is where one may induce more inner disfiguration—Bedrij likens it to heating a sewer without first cleaning it. This may be when one can take on a "spaced-out" appearance.

It is OK for me to learn about management principles and strategies, but I just can't get into this "touchy/feely stuff"—it's just not me.

I can indeed appreciate your need for intellectual understanding and controlled behavior. But the idea that "soft skills" are "touchy/feely" comes from a one-tier consciousness. If we have only the intellect-body-emotions to work with, and we are called to develop, for example, relational skills that are not primarily intellectual, then we think they must be bodily (touchy) or emotional (feely). But in deeper awareness, we live and work out of a two-tier consciousness (intellect-body-feelings—deep inner core) and refer to being at the service of the empowering, inner core energy. Such energy is pure positive power within; we can connect it however we need, to the intellect, to the body, or to the emotions; we can also simply relax and enjoy the inner presence. Consider doing an IT task i) with inner enthusiasm and ii) with apprehension or anger; the intellect is thinking the same thoughts about the technical problem in each case, but underneath this intellectual activity, there is a different, underlying energy. We try to place the most positive underlying energy possible under our intellects and emotions when we work in IT from a two-tier consciousness. Working from the inner being is a joy; everything flows, including capacity to relate, then we don't have to force ourselves into unnatural touchy or feely actions—we can really be ourselves and no one but ourselves (our truest selves, at that).

OK. Maybe, in theory, there is something to it. But, in practice, especially in IT, isn't it simply too way out? Can we really get to a place where we can apply it?

If I didn't believe in or hadn't experienced the capacity to work from one's deepest self (in PRH terms, "to live one's talents"), I would not have taken the effort at such detailed presentation of this psychological dimension. This will not be a "direct cutover" and will take time. Psychological awareness in IT will likely undergo growth stages as postulated in a subsequent chapter. Personality and cognitive awareness will likely come first and take several years to "catch on." However, individual stories, coming particularly from NF types, will be noticed, especially if significant impact on effectiveness, productivity, or stress reduction results. It will take time, but perhaps this is an initial blueprint for some new "happening."

CONCLUSION

In conclusion, we have examined a dimension of the human person that gives solidity, rootedness, and empowerment to the intellect and personality. We have looked at the possibility of functioning in IT from a lived experience of "personal wholeness," which is possible after an inner opening to another level of awareness.

In the words of Joseph Weizenbaum:

*"If the teacher, if anyone, is to be an example of a whole person to others, he must first strive to be a whole person. Without the **courage to confront one's inner** as well as outer worlds, such wholeness is impossible to achieve. **Instrumental reason alone cannot lead to it.** And there, precisely is a crucial difference between man and machine" [bold added].*

We can then ask: Will the IT professional of the 21st century see and experience him/herself as:

1. an intellect, trying to solve a problem; or
2. a coordinated, synchronized "human" "being" using the intellect that receives, channels, and expresses the deeply positive awareness and energy of the Inner Core?

A systematic shift in consciousness from i) *using* one's talents to ii) *living* one's talents may indeed involve "personal re-engineering"!

REFERENCES

Bedrij, O. (1977). *ONE*. San Francisco, CA: Strawberry Hill Press.

Borysenko, J. (1990). *Guilt is the teacher, love is the lesson.* Warner.

Carrington, P. et al. (1980). The use of meditation/relaxation techniques for the management of stress in a working population. *Journal of Occupational Medicine*, 2(4).

Couger, J.D., & Zawacki, R. (1980). *Motivating and managing computer personnel.* John Wiley & Sons.

Damasio, A. (1994). *Descartes' error.* New York: Avon Books.

DeMello, A. (1990). *Awareness.* London: Fount.

Diamond, J. (1994). *Your body doesn't lie.* Warner.

Dyer, W. (1995). *Your sacred self.* Harper Paperbacks.

Friedman, M., & Rosenman, R. (1974). *Type A behavior and your heart.* New York: Fawcett Crest.

Gardner, H. (1983). *Frames of mind: The theory of multiple intelligences* (2nd edition). New York: Fontana.

Goleman, D. (1995). *Emotional intelligence.* Bloomsbury.

Goleman, D. (1998). *Working with emotional intelligence.* Bloomsbury.

Helminiak, D. (1987). *Spiritual development: An interdisciplinary study.* Loyola.

Helminiak, D. (1998). *Religion and the human sciences: An approach via spirituality.* New York: SUNY Press.

Hoffman, B. (1988). *No one is to blame.* Recycling Books.

Lesiuk, T.L. (2003). *The effect of music listening on the quality-of-work of computer information systems developers.* PhD Dissertation, Faculty of Graduate Studies, University of Western Ontario, Canada.

Longenecker, C., Simonetti, J., & Mulias, M. (1996). Survival skills for the information system professional. *Information Systems Management*, (Spring), 71-77.

McGraw, P. (2001). *Self matters.* New York: The Free Press.

O'Brien, J. (1998). *Introduction to information systems—essentials for the Internetworked Enterprise.* McGraw-Hill.

Orsborn, C. (1992). *Inner excellence.* San Rafael, CA: New World Library.

Pentecost, K. (2003). XP and emotional intelligence. *Cutter IT Journal*, 16(2), 5-11.

PRH International. (1997). *Persons and their growth.*

Rogers, C. (1961). *On becoming a person.* Constable.

Siegel, B. (1989). *Peace, love and healing.* New York: Harper & Row.

Stokes, S. (1996). A line in the sand: Maintaining work-life balance. *Information Systems Management,* 13(2), 83-88.

Thierauf, R. (1988). *New directions in MIS management: A guide for the 1990s.* Quorum Books.

Weizenbaum, J. (1976). *Computer power and human reason: From judgment to calculation.* W.H. Freeman.

Yourdon, E. (2002). *Cutter IT Journal,* (December).

Zukav, G. (1990). *The seat of the soul.* Fireside.

Zukav, G., & Francis, L. (2001). *The heart of the soul.* New York: Simon & Schuster.

Conclusion:
Part I

In this major part, I have tried to present an introduction to several significant psychological realities, the awareness of which, I believe, may indeed have a significantly positive impact on the work of IT professionals. I have attempted to include enough material on each psychological dimension so as to provide fundamental knowledge rather than a generally descriptive overview. I have tried to expose several tools for carrying out a "systems analysis" on one's inner functionings.

We have examined how we tend to function and where we tend to focus; how we tend to think, learn, and create; and how deeply aware we are of our essential human identity. Some groundwork has thus been made for the beginnings of inner transformation to a more effective, resilient, and empowered person, capable of carrying out IT work with "emotional intelligence."

A central vision promoted here has been that of a two-tier consciousness, the "human" connected to the "being." Such an actualized connection is what, in fact, makes "emotional intelligence" possible. It enables truthful appreciation of one's own feelings and that of others in different situations. The connection also prompts wise responses to perceived emotions, as guided by the deepest self, responses that may yield "win-win" situations.

When one becomes skilled at not only noticing feelings, but also being aware of personality and cognitive functionings, and responds more wisely because of such awareness, one can be said to be operating with "*enhanced emotional intelligence*" (EEI). We can attempt to define such EEI in an IT context as:

...ability to respond more effectively in given work situations based not only on technical/managerial knowledge, but on an awareness of the psychological functionings of self and others on the levels of intellect, emotion, and deep inner self (spirit).

This part has attempted to provide an initial base for establishing a coherent framework for understanding, promoting, and developing enhanced emotional intelligence in the IT profession. As an initial base, it encourages psychological awareness and understanding on several dimensions only. It does, however, initiate and encourage further efforts for evolution into a more comprehensive, applicable base for psychological understanding in IT. For example, techniques and concepts such as "neuro-linguistic programming (NLP)" and "focusing" can, at some point, be introduced.

It is hoped that material in this part does act as a catalyst in providing greater psychological awareness among IT workers individually, among IT organizations, and among interested researchers. While Part I provided a basic orientation, Part II follows with an overview of how such an orientation can be applied systematically in IT on an individual and organizational basis.

Part II:
Making IT Work

The first and major part of this book focused on introducing several psychological dimensions, and, intermittently, on pointing out how explicit awareness of such dimensions in oneself and persons with whom one works can assist in IT work. Ultimately, the first part laid the foundation for "enhanced emotional intelligence," which can be described as "ability to respond more effectively in given work situations based not only on technical/managerial knowledge, but on an awareness of the psychological functionings of self and others on the levels of intellect, emotion, and deep inner self (spirit)." It developed this concept by not only encouraging awareness of emotions during specific work incidents, but providing insights as to possible roots of such emotions (as in the Enneagram) and promoting a way to transform and "ground" emotions for optimal functioning (as in the deepest inner self). As well, it introduced psychological aspects of personality and cognitive styles, which, if applied judiciously, can beget positive emotion and diminish inner conflict. Part I also aims to encourage IT workers, higher IT management, HR persons involved with IT workers, and IS researchers to consider more seriously and centrally both the topics presented and related areas that have not been included. Hopefully, if, over time, IT workers assimilate and integrate such new awareness, they will ultimately work more wisely.

This part attempts to reference coherently and to encourage specifically *direct application* of such enhanced wisdom to the IT environment. It concerns the *management* of the psychological factors. Chapter V presents an overview of the benefits of EEI in some key areas of concern to IT work.

Chapter VI looks at development of psychological awareness in IT from an organizational perspective. Finally, Chapter VII proposes specific initial steps that can be taken by different parties involved to allow the ideas heretofore presented to take on form within IT.

However, as it will be evident, this part is only an *initial attempt* at delineating application possibilities. It is intended to orient, arouse curiosity, and motivate specific attempts at application of the psychological factors. With interest and adequate feedback from the "IT trenches," it is quite possible that, in time, a separate book will evolve from this second part. For now, this part can serve as an "appetizer," a foretaste of real and impactful possibilities. Most importantly, it is hoped that this part, with its applied orientation, will indeed begin to generate new experiential realities in IT work, realities that energize and empower.

Chapter V

Areas of
IT Application

INTRODUCTION

A natural question in the minds of many readers is, "How is all this *really* going to influence daily IT work?" This chapter attempts to provide some indication in an organized, albeit overview form. Thus, when discovering areas of relevance to IT management, such as teamwork, end-user relationships, motivation, change management, etc., the goal is not to present comprehensive coverage of each such area. The aim, more modestly, is to provide an initial attempt to show *how* ideas presented in Part I can specifically influence such areas of relevance. It is hoped to demonstrate potential effects of working with "enhanced emotional intelligence" as outlined earlier and, thus, to arouse further motivation for application of such a framework in specific areas outlined in this chapter.

IS TEAMWORK

A considerable amount of system development is accomplished through the use of project teams. Indeed, much has been written in management and IS literature about effects of team synergy. However, what would characterize an "emotionally intelligent" system team? How could it *apply* the considerations from Part I?

Team Formation

In the book, *The Wisdom of Teams*, Katzenbach and Smith (1999) define a team as "a small number of people with complementary skills who are committed to a common purpose, performance goals, and approach for which they hold themselves mutually accountable." At the start, EEI can assist in the first stage, *team formation*. Choosing team members, not only by technical competence but also by considering anticipated psychodynamics, is indeed recommended. Whether it be through formal MBTI profiles or a general, but accurate appreciation of at least the temperament types (NT, NF, SJ, and SP) and their strengths, a personality type focus can eliminate narrowness of approach and lack of balance.

A team organizer will know the nature of the project (e.g., degree of innovativeness, degree of interaction with users, amount of detailed coding required, etc.). He will be aware of the strengths and weaknesses of each temperament. To support his impressions, he may actually interview prospective team members and ask them work-related (and then specific experience-related) questions that would confirm their temperamental traits and preferred modes of operating. He can try to elicit views of how a person with a particular temperament would interact with another specific type. Such consideration could form one major variable in a (albeit informal) decision support model as to whom to select for the team.

Task Assignment/Execution

After the team is chosen, *task assignment* is another main requirement. There will likely be subgroups on the team responsible for different project aspects (e.g., data model, process model, GUI design). Consideration of how personality traits and thinking styles will complement can assist greatly. For example Patricia Ferdinandi's proposed set of personality strengths for various system development tasks (Table 10) can be assessed and then tested out. When explaining to a team member *why* she was chosen for a particular task (e.g., user interface design), personality traits, formally assessed or perceived, can be included in the discussion.

After task *assignment* comes task *execution*. Here, the team (and possibly its subgroups) is at work. Specific awareness of ones MBTI personality preferences and those of co-workers can indeed be valuable. Persons can get to appreciate each other's strengths, and realize when and from whom to ask for help. Initially, awareness can be more rudimentary, focusing individually on each of the four dimensions: Introversion/Extraversion, Intuition/Sensing, Thinking/Feeling, and Judging (structured)/Perceiving (open-ended). For example, a sensing, detailed coder may ask advice on different possibilities for data capture from the possibility-oriented intuitive. Also, awareness of thinking styles can provide similar advantages—one person designs the overall system functionality, the other, the underlying structure.

In time, as team members learn through several successful projects to appreciate the value of personality awareness, they might go deeper to consider, for example, *orders of preference* (Table 7) for each of the 16 MBTI types. For example for an INTJ, intuition "kicks-in" first and sensing last, but for an ISTJ, sensing activates first and intuition last among the four functions of intuition, sensing, thinking, and feeling. Persons with such an advanced awareness bolstered by considerable experience could no doubt communicate much more functionally and synergistically. For example, when met with a system requirement, the person starting from intuition would likely see and express something quite different than the person starting with sensing. Instead of arguing and defensiveness, or at least fairly ineffective communication, psychologically aware system developers would indeed appreciate and welcome each other's points of view and would strive towards a "coherent whole" by combining effectively individual perspectives. Empirical research and extensive dissemination of noteworthy results to the professionals as well as other academics would further advance the application of such enhanced self-awareness.

While team members can certainly benefit from awareness of their strengths and complementarities, they are also advised to pay significant, direct attention to their own emotions and those of others. Such emotions can motivate, energize, and bond, or conversely, de-energize and break apart intended team cohesiveness. A starting point for many may be one's Enneagram awareness, which identifies, specifically, an underlying emotion (and a resulting "coping pattern") for each of the nine types. After having reorganized and acknowledged one's operational "compulsion," one can strive consciously to move towards one's intended *direction for integration* (Figure 1). Furthermore, when team members trustingly admit to each other their common emotional traps and try to transcend them openly (while asking for patience and understanding from others), considerable pent-up psychological energy may be freed up for constructive use. For example, a Six, who is driven to rely on loyalty to the "tried and true," may have a lot of difficulty accepting, emotionally, a new development approach/methodology. Yet his loyalty to the team causes him to "act as if" he is really "in sync" with the new mode of work (he is moving from Six to Three, his direction of *dis*integration). By admitting all this genuinely, without self-judgment, he now tries to move to Nine and attempts to be "at peace" with his unfamiliar-ity—he is able, slowly, to let go of his excessive attachment (loyalism) to old, habitual work modes.

The most likely last avenue of inner development, but with extremely significant potential, is *growth in the deepest inner self*. Here, as we have seen, a person evacuates inner blockages to allow access to a pure core en-ergy of wisdom, truth, creativity, peace, and joy. Connecting this spiritual energy to one's emotions, intuition, and intellect, and making it the "master" of one's inner life provides emotional resiliency, courage, and capacity for wise decisions. It is the deep inner core energy that makes the emotions "intelligent."

In teamwork, a team that, in addition to personality and cognitive con-sciousness, is significantly able to work out of the deepest self will undoubt-edly be the most functional, efficient, effective, and energized team that one could imagine. Several points are in order here. A desirable team has cohesion, but not "fusion" (which could lead to "groupthink"). People are free to be fully themselves while deeply aware of their own innate dignity and that of others. Disagreements are handled constructively since the creative, positive energy within each member transcends the disputes and automatically places them in perspective. Also, a key result of being deeply connected inwardly is that one is able to *respond* to a difficult situation (such as a moved-up dead-line) from inner wisdom, rather than *reacting* from one's blocked emotion. Reactions (e.g., in anger, fear, cynicism, blame) drain energy—responses empower from within.

Also many people, perhaps more so those who may rely heavily on structure for significant emotional security, work with significant *expecta-tions*. They expect specific things to happen in a specific way (e.g., for their program to work, for the user to be satisfied). With the uncertainties of system development, unmet expectations can certainly be very emotionally draining. When one is inwardly connected, however, one *interacts* with the structure but is *not attached* to it; the security comes gratuitously—from deep energy within. Thus one can approach system development teamwork with *expec-tancy*, i.e., a general optimism that one has enough vivifying energy within to deal as best one can with whatever comes up, and one will live comfortably with however one was able to deal with the difficulties.

Communication/High Performance

An inward rootedness gives a flavor of *authenticity* to interpersonal com-munications. Trust follows more naturally, as do a sense of vision, purpose, and a genuine "team spirit." Such a team is indeed at an advanced level of enhanced emotional intelligence. Although one may think of such synergy as utopian, in light of practical evidence from sources such as Siegel, Borysenko, Zukav and Francis, and Bedrij (who certainly knows about computers and software) regarding the power from the inner self, such synergy can at least be systematically, yet realistically approached.

Steve McConnell, in *Rapid Development* (1996), identifies the following characteristics of a "high performance team":

- Shared, elevating vision or goal
- Sense of team identity
- Results-driven structure
- Competent team members
- Commitment to the team
- Mutual trust
- Interdependence among team members
- Effective communication
- A sense of autonomy
- A sense of empowerment
- Small team size
- High level of enjoyment

It is not hard to see that such a team is realized mostly through emotional intelligence, through rootedness of its members in their deepest inner self. In a sense, *such a team is spiritual*. However, it is postulated that unless one can achieve, through inner growth work as outlined in Chapter IV, a connection to his/her inner dynamism, such high-performance characteristics can be developed only quite partially. Yes, one can change one's principles and beliefs about teams; one can modify one's behavior through insight, exercises, and willpower; one can even initiate some emotional bonding with others—but as long as these changes happen with a one-tier consciousness (intellect, emotions, body), they will be only partially effective.

We can now recall the statement, "Our humanity is extremely underdeveloped in the area of in-depth riches." Full realization (with the inner eyes) of this statement and the resulting commitment to full growth, to "personal wholeness," will no doubt realize considerably the above listed characteristics among software development teams.

END-USER RELATIONSHIPS

This is indeed a major topic when discussing the success of system development efforts, particularly in the near future with many e-commerce, multimedia systems for competitive advantage being the focus of development. In such situations in particular, understanding user requirements and eliciting user assessment throughout development activity and adequate user training are undeniable necessities. Here, a developer with EEI will have distinct advantages.

Firstly, an understanding of the four temperaments and thinking styles, as well as experience in how each type tends to talk about system requirements, is a very useful skill. NTs may tend to talk about system functionality in abstract terms; NFs may focus on how the system will assist users/customers; SJs may get into the "nuts-and-bolts" of specific outputs, and SPs may relate to functionality, but from a practical viewpoint. On the part of the analyst, knowing one's own temperament biases may indeed improve communications (e.g., What questions might an SJ ask an NT when determining system outputs?). Such skills, however, would likely take some time to develop and could well be reinforced by "case study" training (e.g., through videos or CDs).

Since considerable apprehension on the part of the users often relates to a perceived lack of "humanity" on the part of technical developers, growth in the deepest self would enable one to demonstrate his authentic humanity at its best. Working from the center deep within, an analyst could relate to a user with respect, openness, flexibility, and a felt sense of genuine interest to assist the user. Trust would follow as a corollary, opening valuable communication channels. In time, the analyst could listen attentively, while being aware of the other's communication style with her intuition, and offering impactful presence by awareness of her own inner depth. Such a synergistic awareness on the part of the analyst, along with a resulting capacity to relate, would be invaluable in the development of complex, yet creative, influential systems.

Research, perhaps in the area of "Psycho-informatics" on effectiveness of analyst-user relationships, where the analysts possess enhanced emotional intelligence as defined in this book, could provide very valuable insight. Until now, research has confirmed that users' attitudes towards an information system are considered to be an important indicator of the quality of the implementation of the system. Thus, according to Kaiser and Sirivasan (1982), "It is critical that the systems designers have attitudes about the system and its environment that are compatible (not necessarily concurrent) with those of the user." As well, these researchers have found that "both users and systems staff do indeed perceive user-analyst communication as an integral part of systems development and implementation."

Thus, a team of analysts would be well-advised to each develop an inner "attitude barometer" with respect to user relationships. Such a skill may indeed extend beyond cognitive/intuitive perception and into the "deeply felt sense" of reality as arising from the deepest inner self. Such a locus of control, not only internal to the analyst, but found in the deepest, trans-intellectual part within the analyst, should also provide the enhanced emotional intelligence necessary to foster fruitful user-analyst communication.

This approach could operationalize the advice provided by researchers Newman and Robey (1992):

"Attention should be paid to individual encounters between analysts and users [and] the project leader could manage the process by watching for signs of progress or degeneration and then provoking encounters for the purpose of raising issues."

The researchers have presented a social-process model of user-analyst relationships which, they proposed, could be of use to the practitioner in predicting the consequences of actions taken during information system development. Perhaps the degree of psychological awareness and emotional intelligence, as promoted in Part I, could become specific components of an extended model.

In other research, Hartwick and Barki (1994) have noted a number of psychological variables (e.g., involvement, attitudes, and subjective norms) that intervene and mediate the relationship between user participation and system use. With enhanced psychological awareness, not just generally, but involving specific factors such as awareness of user temperament and cognitive style, perception of user's Enneagram type, and degree of deep inner-centeredness of the analyst, such a research approach could, no doubt, be extended.

Alavi and Joachimstaler (1992) had developed a framework suggesting that individual user factors that are most relevant to acceptance are cognitive style, personality (traits include need for achievement, degree of defensiveness, locus of control, dogmatism, and risk taking), demographics (variables include age, education), and user-situational variables (training, experience). It is indeed noteworthy that the identified personality traits can be significantly "normalized" by a two-tier consciousness, i.e., a significant connection to the deepest inner self.

Another area of psychological awareness in relation to users is that of *learning style*, as presented in Chapter III. Preparation of training programs for new systems implementation can involve consideration of how people learn best. Also, emotional barriers to learning on the part of the users could be discovered and addressed. Considerable research in this area would provide very practical application.

Thus, awareness of psychological factors can be of immeasurable value for IS workers in dealing with end-users. However, this awareness would make a significant impact only after considerable self-development along the dimensions proposed in this book. Such development could be catalyzed by informative websites and appropriately constructed learning materials (case studies, instructional videos, etc.). It is one thing to tell a developer that her user might be an intuitive or a sensor, but it is another to show a video, in a specific system context, on *how* an interaction with each type of user might look and why.

CONFLICT MANAGEMENT

Barki and Hartwick (1994) have also studied *conflict* during systems development. They note that "conflict has come to be seen as a dynamic process, something to be managed" (Kolb & Sheppard, 1985), and that it can be seen as an opportunity for growth and change. Information system development involves multiple interdependent parties who often have differing interests and goals. During the development process, each party may likely behave in a way that furthers its goals. When a party's actions interferes with other parties' goals, conflicts occur.

Barki and Hartwick further note that:

"conflict, as well as other interpersonal processes such as participation, influence, and conflict resolution can be studied at the individual or group level of analysis."

They then quote Robey et al. (1989):

"The individual level of analysis is appropriate for examining the psychological experiences of participation in a group process, influence over others, and engagement in conflict and its resolution."

The influence of specific psychological awareness factors on such psychological experiences of participation could indeed provide fruitful results, both theoretically and practically.

Development in psychological awareness can indeed make a system developer much more equipped to handle conflict constructively. Conflict, whether with users, higher management, or other system developers, will escalate as each party adopts a defensive and inflexible position, fearing being ignored or diminished. Awareness of different perceptions arising from different orders of preference for different MBTI personalities, for example, would undoubtedly cause an astute analyst to adopt a more flexible mindset in key negotiations. Such a person would indeed appreciate that one would notice and defend a different position if his thoughts "kicked-in" first than if his feelings and then intuition "kicked-in" at the start.

As well, an analytic cognitive style may indeed initiate different perceptions and concerns than the heuristic style. A skilled developer could also notice likely hidden (or, not so hidden) emotions in the course of conflict and link them to an opponent's likely Enneagram type. Through insightful communication, such a developer could lead the opponent towards his/her *direction of integration*, diffusing considerable negative energy. Also, a presence rooted in the deep self can, by its nature, keep conflicts at a constructive level amenable to resolution. Authenticity and emotional security can allow for "radical honesty" and an intelligent exchange of differences in perception or motivation.

OTHER IT MANAGEMENT ISSUES

Apart from teamwork, user relationships, and conflict resolution, there are a number of other management areas in IT, on varying levels, where psychological awareness could indeed have significant impact.

Project Management

IS is the most ongoing, project-oriented area of a business. Thus, strong project management skills are indeed a requirement. Aside from task management, mentioned while discussing teamwork, one main area here is *scheduling*. In estimating times for specific tasks, an experienced PM could anticipate extra time that may be needed by particular staff members who may require additional emotional adjustment. For example, a traditionalist SJ type is asked to use a new development tool. While it might be in order to add to the expected time, normally five days for familiarization with the necessary features, an extra two days may be in order to allow for this developer's inner adjustment to this change. If one kept statistics on the relationships between temperament type and time taken to carry out tasks that are "natural" and "unnatural" to the type, more accurate scheduling may well result.

Also, the PM can become explicitly aware of his/her own emotional biases regarding scheduling. A "Loyalist-6" may be so preoccupied with meeting deadlines preferred by a user that he may disregard obvious negative effects on the morale and energy of his staff. Robert Glass (1997) reported, "About 40% of all software errors have been found to be caused by stress; those errors could have been avoided by appropriate scheduling." He also presented a finding that, when schedule pressure was extreme, defects in a released product increased four-fold.

A project manager, therefore, must have an astute inner sense of "what is reasonable." He must also be capable of noticing employees who are working with "pressure addiction." Such persons thrive on the "high" that they experience when rushing to meet deadlines. The danger here is that such a pace cannot be maintained for a long time and eventual burnout will be the likely result. Pressure addiction should be recognized as a basically unhealthy coping mechanism, largely resulting from unexpressed feelings and poor access to the revitalizing inner self.

Another PM skill is *motivation*. An area of psychological awareness involves realizing what are the main motivating factors for each of the subordinates. Such information can be gleaned during authentic, informal communication. A useful insight (and another research challenge) would be to relate temperament and underlying emotions to specific motivating factors. Steve McConnell presents an ordered list of motivators for programmer analysts extracted from research by Boehm (1981) and Fitz-Enz (1978). The top five motivating factors are achievement, possibility for growth, the work itself, time for a personal life, and opportunity for technical supervision. In this light, if psychological factors were accepted as a specific aspect of work life, possibility for growth would be satisfied along a new dimension. A sense of achievement and the work itself can be bolstered by assigning persons to tasks that fit their innate strengths. Opportunity for supervision can be enhanced by explicit awareness of not only technical, but also psychological factors.

In motivation and supervision, there will come a time for the PM to be firm (e.g., regarding deadlines, standards, or even persistent user changes to established requirements). For many, being firm may be unnatural; firmness, then, is equated with rigidity. Faced with an inner fear of adverse reaction, a PM may tense up, repress the fear, and pronounce an edict. Such is firmness in a defensive mode; such firmness often provokes adverse reaction (perhaps not directly expressed). Yet, for one deeply connected to the "truth within," true firmness is indeed natural. He can insist on what is necessary from a place of inner authority, while simultaneously respecting and inwardly acknowledging the dignity of the receiving party. Such firmness will be, in all likelihood, very differently received. A person living on two tiers has the ability to be indeed emotionally close, yet inwardly free.

Another PM issue is that of *control*. According to researchers Henderson and Lee of the School of Management at Boston University, "The control relationship between the project manager and the team members is central to effective performance in…an IS project-oriented task environment." Others have described control as a process that can involve the monitoring of both behaviors and outcomes. Behavior control can include role clarifications, work (task) assignment, and procedure specification (enforcing rules, procedures, and work methods). Outcome control can be achieved by arriving at specific milestones. In their research, Henderson and Lee (1992) found that when behavior-based control and outcome-based control were compared, the behavior-based control was more significantly related to all performance variables. They then concluded:

"The results highlight the need for IS project managers to have skills in exerting behavior control. And yet, follow-up interviews indicate a serious limitation in this respect… Our results suggest a need for the project manager to understand the processes and dynamics of the design process itself."

Jim McCarthy, in his book, *Dynamics of Software Development* (1995), seems to echo the above remarks:

*"Leaders, good ones at any rate, are expected to provide a vision to their followers. In my opinion, **a leader's empathic perception of the psychological state of his or her team is the beginning of what we call vision**. If the leader can then resonate with the team's complex emotional state—identify with it, articulate it, and give the whole constellation of feeling and thought a visible, concrete reality in his or her own personal voice or gesture—the boundaries among individual team members and the leader will collapse…Empathy will be established: the leader and the team will feel and know as one, giving voice and identity to what was an incoherent psychological community substrate.*

And it will feel good to the team to be understood and to understand... Without empathy, vision is hollow, an ersatz vision that might fulfill a requirement that some words appear on a slide, but won't provide the visceral motivation that inspires a team to greatness" [bold added].

While reading McCarthy's eloquent and fitting words, it is not difficult to realize that his "empathy" reflects strongly a connection to the deepest ("visceral") inner self, or in other words, emotional intelligence supreme, or authentic spirituality.

Another psychological requirement of project managers is that of *cognitive complexity*. Gina Green (1997) examined the impact of such complexity on the effectiveness of project leaders. Cognitive complexity studies humans as information processors. Cognitive differentiation is an individual's ability to dissect information into smaller units; integration is the ability to combine smaller units of information into a whole; the latter is often deemed the more important component of cognitive complexity. Green states that "the link between cognitive complexity and project leadership performance should encourage MIS managers to seek actively ways to improve levels of cognitive complexity of their project leaders." Suggested approaches include training leaders to view problems from different perspectives and be "cognitively flexible" in implementation of strategies. Clearly, such prescription would necessitate awareness of one's cognitive style, and within it, preferred ways of integrating information. Also emotional resistance to developing cognitive complexity could be systematically researched.

In time, it is likely that psychological awareness and "emotional intelligence" will be seen as required, not optional characteristics of most IS project managers, particularly those in senior/influential roles. Jim McCarthy's remarks "stand tall" and such realizations cannot be ignored for long. However, the IS field cannot just expect that project managers will "somehow" develop such competencies on their own. To this end, significant development materials will need to be produced and made part of IS project managers' certification processes.

However, such a concerted effort to embed psychological awareness into the project managers' skill set will need a "kick-start." A website for IS project managers with information, application cases. and success stories would be very desirable. Support for such a site from official PM bodies would serve to legitimize the effort. The production of extensive training and development materials (many likely multimedia based) for psychological awareness among project managers can be motivating and creative. Yet, the question remains, "Who will do it?"

It would, thus, serve the interest of the entire IS profession as well as the MIS research community if a "Psycho-informatics Center" were established jointly by industry and academe. Such a center could centralize the extensive development work needed to provide the profession with thorough, practical educational and development tools/methodologies to enable widespread, competent application of psychological factors to daily IS work. It is hoped that the feasibility of creation of such a center could be discussed in the proposed website related to this book that should follow the book's publication.

Specific Development Approaches - JAD and XP

It is worth addressing the issue of psychological factors in specific system development approaches, e.g., Joint Application Development (JAD) and the more recent Extreme Programming (XP). JAD is a requirements definition and user-intensive design approach/methodology in which end-users, executives, and systems developers attend intense meetings (usually off-site) to work out the details of a new system. It shortens the normal time needed for requirements definition, and it makes it possible to gather such requirements more effectively. It can best be combined with rapid-development and prototyping tools.

The JAD room is specifically structured and includes people with specific roles: session leader (a pivotal player), executive sponsor, end-user representative, developer(s), scribe, and resource specialist(s). The complete JAD group should usually not exceed eight people. During JAD, end-users talk with analysts and each other about desired system features.

Such an environment can certainly benefit from considerable psychological awareness on the part of the systems developers. Recognizing and responding appropriately and immediately to different thinking styles can certainly decrease miscommunication. Authentic communication centered in one's deep inner core can motivate cooperation and remove communication barriers. Inner connectedness can also provide emotional rejuvenation during intense planning and design efforts.

Extreme Programming is a relatively new approach to software development. It is based on values of simplicity, communication, feedback, and courage. The development team forms around a business representative (customer), who is always available. The team produces software in a series of small, fully integrated releases that passes all the tests that the customer has defined. Extreme programmers work in pairs at the same computer. One partner provides strategic steps while the other develops tactical workings.

Programmers are moved around, often from pair to pair, and they all code in a consistent style. Unit testing is viewed as critical. There is collective code ownership. XP teams may build software in two-week "iterations," delivering running, useful software at the end of each iteration. XP focuses on delivering business value in every iteration. XP teams build software to a simple design. The team keeps the system fully integrated at all times. XP teams build multiple times per day.

An immediate place for psychological awareness in XP is in the programming pairs. Explicit realization of each other's cognitive styles can make collaboration more efficient, productive, and satisfying. To quote Ben Konitz in "XP and the Cognitive Divide" (2003):

"XP exploits the fact that different people think differently...the rhythm of pair programming consists of each person constantly showing his or her partner things that to one person seem obvious, but might have taken days to occur to the other—or might never occur."

Also, there is less formalism within XP, as reality itself "guides you to what can be done and even what should be done through constant, concrete interaction." Might it be true that, for example, certain Enneagram types are more suited to the XP environment, at least initially, than others? Flexibility is a requirement in XP. Would the rigid, Perfectionist-1 type be frustrated by this? Without an Enneagram growth path, might he not lose energy in fruitless self-analysis (the direction of disintegration for a One)? In any case, one gains significant psychological capacity for adjusting to feedback and for courageous communication when guided by his deepest self (thus, being *conscious* of his profound stability rather than *self-conscious* of perceived shortcomings).

Lowell Lindstrom and Kent Beck (the originator of XP) report that from their experience, "Introducing teams to XP causes enormous pain and dislocation." With a deeper inner connection pointing to a truly stabilizing energy, such dislocation can be mitigated. Beck also dismisses the idea that IT work is "some Vulcanic world of pure rationality" (this seems to echo Dr. Damasio's assertion in *Descartes' Error)* and declares:

*"To be successful **we must learn and grow our whole selves**" [bold added].*

Change Management

With XP as an example of significant, specific change, we can consider briefly the role of psychological awareness in managing change in general within IT, an area of constant change. To quote Stewart Stokes in "Coping with Change at the Top" (*Information Systems Management*, Winter 1996):

"During times of rapid change, IS professionals need what is most lacking: honesty and communication."

It would appear that computing initially appealed to IS workers precisely because it avoided the uncertainty of human relationships. With a more predictable IS work environment as in the early COBOL years, a person could maintain significant inner emotional blockage and still thrive at work on one's peripheral talents (those to "do" rather than those to simply "be"). Today, the constant change has eroded the emotional buffer; much more inner energy is needed to assimilate change and deal with significant ambiguity. Such inner energy is plentifully available, but must be "tapped into" by evacuating inner barriers, many of which arose by a lack of a fully nurturing emotional environment during one's earlier years. Such an inner experiential connection provides "no fear, only love, peace, wisdom, interconnectedness and an overwhelming sense of…safety that abides in the now" (Joan Borysenko).

Change management consultant Diana Larsen in "Embracing Change" (2003) notes:

*"Change agents and those involved with change **who have the personal and social skills of emotional intelligence will show more resilience** in the face of a shifting environment" [bold added].*

In response, quoting PRH:

"At the level of the Being, there is everything that is needed for persons to create harmonious relations with those around them and to cooperate effectively in a common task. There is everything needed so as not to capsize under the pressure of the hard blows of life {change} and to restore a solid footing within oneself so as to carry on with a firm step."

However:

"the chief executives {and MIS managers}…must choose this integral growth for themselves and experience its benefits."

Inner growth-in-being is a definite answer to the quest for resiliency.
However, it does require systematic work and, yes, even more change.

Relationship with Higher Management

Another central issue in IT management, particularity at the CIO level, concerns relationships with higher management. An effective CIO needs to have respect of and influence with higher organizational management. Human factors, among them psychological awareness, can play a significant role indeed in the potential impact of the CIO. While personality type considerations (often Myers-Briggs) and possibly cognitive styles may be included in communication training programs that CIOs might undergo, the wisdom of the Enneagram and of deeper inner rootedness has not yet made its rightful impact.

We can recall, from Chapter II, the influence that an integrated Two type (Helper) can maintain in an organization and the impact of an appropriately power-seeking Eight type. Awareness of Enneagram motivations and directions of integration can be very useful for the CIO wishing to maintain influence. Also, *much can be accomplished in terms of impact and influence simply by "presence."* This does not, of course, refer simply to physical presence or intellectual acknowledgment, but an ability to be wholly open, listening and engaging with enhanced sense of inner being, authenticity. and emotional intelligence. **To be truly "present," one must "grow one's whole self" as suggested by Kent Beck.** Thus, an enlightened CIO can learn to ask herself, *"From where within me am I responding to the issues in this meeting?"* and then to engage all available inner resources for optimum effect.

A Call to "Personal Wholeness"

Dan Ariely is an endowed professor of behavioral economics at MIT's Sloan School of Management. As a behavioral economist, Ariely studies how people make decisions in real life and why their decisions often deviate from classic economic models. Ariely has studied how people's experiences of physical and psychological pain affect their decisions. This can be related to the question of why good CIOs make bad decisions. According to PRH, Hoffman, and similar growth movements, psychological pain, often lodged in the subconscious, arises when the true, deep self is not attended to sufficiently in one's developing years. It is such pain or inner blockage that distorts one's capacity to act with the wisdom lodged within one's depth. Again, a reason to "grow the whole self."

A call to psychological awareness within IT culminates by a call to "personal wholeness," which can be defined as "connectedness of the mind, body, and feeling to the inner being, such that the energy of the being can express itself through the other three parts." In such a state, a person is "fully himself, only himself, and no one but himself (especially the deepest self)." He/she is in inner harmony and self-actualized. PRH and Hoffman have expounded on the benefits of such psychological integration; Weizenbaum points to such wholeness as the fundamental difference between man and machine; Beck, *from within the computing profession,* calls for such development. Has the time indeed come for a large-scale effort for psychological emancipation in the IT field?

STRESS MANAGEMENT

Intuitively, few would argue with the supposition that considerable work in IT at this time is significantly stressful. Indeed, there is both initial exploratory survey as well as anecdotal evidence supporting this claim.

IT workers have reported perceived causes of stress as being increased demands on IT for more complex systems, rising workloads (the need to do more with less), rapid change (in technologies, methodologies, and organizational structures), communication problems, and unrealistic expectations from users. Change and uncertainty have undoubtedly characterized the working climate for today's IS professional.

One recent survey of 1,400 CIOs found the top three sources of stress for IS workers to be: i) increasing workloads (55%), ii) office politics (24%), and iii) imbalance between work and personal life (12%). Another such survey showed the three greatest IT staff concerns among 290 IT executives to be: i) demanding workloads and resulting burnout, ii) retaining key IT skill sets, and iii) low morale. Half of those surveyed described the stress level among their IT staff as high to very high.

In my own research of 191 Canadian IS workers, 53% stated that their job was stressful, 51% reported unreasonable deadlines, 51% noted significant absenteeism due to stress, and 54% approved of a specific effort to combat stress from within the IS profession. Also, those workers who had considerable difficulty in "turning off" from work at the end of the day had significantly higher expectations of an impending health problem.

Since I was concerned that my figures might be somewhat biased (as really stressed workers may not have wanted to fill out a 20-minute questionnaire), I initiated a short, Web-based poll on IT stress and obtained somewhat different results from 250 online responses:

- 72% reported at least five out of seven as to whether their job was stressful (36% at least six out of seven)
- 48% reported at least five for being close to burnout (24% at least six)
- 63% reported at least five for having difficulty stopping to think about work at day's end (50% at least six)
- 44% reported at least five as to work problems keeping them awake at night (32% at least six)
- 50% had reported at least five for having difficulty relaxing (36% at least six)
- 55% reported at least five for the degree of having their sense of worth dependent on the job (24% at least six)

Such results, although exploratory, definitely go along with the perception of significant stress difficulties in IT work. Furthermore, a number of recent researchers and authors have also expressed their concerns. Khosrow-Pour and Culpan (1989-1990) remarked:

"Information processing professionals see change in technology as a prerequisite for their existence, yet the speed of this change can have profound psychological and physiological effects."

In the book, *Technostress* (1984), Craig Brod pointed out that:

"high performance with high technology can exercise a dangerous influence on the human personality...anyone who is constantly working or playing with computers is at risk."

Psychologist Mary Riley, in *Corporate Healing* (1990), points out dysfunctional behaviors arising when:

"high touch has not kept up with high tech."

At this point, we can also consider the caution of Joseph Weizenbaum regarding warm human relationships as an endangered human phenomenon.

The IT world, at this point, has indeed begun to recognize widespread stress among IT workers and the dangers of an impending burnout epidemic. For example, in December 2002, the *Cutter IT Journal* dedicated an entire issue to "Preventing IT Burnout." In it, Guest Editor Ed Yourdon states: "Burnout is still a topic that most senior managers would rather not confront, but it has become so prevalent and severe that some IT organizations have become almost completely dysfunctional."

Is IT stress related to a diminished "emotional intelligence"? If so, how, and to what degree? These are indeed worthy research targets. In 1993, Wastell and Newman remarked:

*"...it is surprising that system development, with its problems and vicissitudes, has not been examined from a stress perspective; indeed **the literature is curiously silent on the general subject of emotional factors in IS development**" [bold added].*

It can even be argued that increased emotional intelligence or greater inner self-awareness may be the most potent weapon against widespread IT work stress.

The aim, in this section, thus, is not to propose or analyze a variety of suggested and mostly indeed valid approaches to preventing and treating IT burnout. The aim, rather, is to examine *how greater psychological awareness in the dimensions presented in Part I can be significantly influential in reducing IT stress.*

Proposed remedies for burnout have often followed traditional stress management recommendations, such as healthier diet, more frequent exercise, and engaging in a more balanced lifestyle. With such an approach, the stress-managing IT worker is still likely to have the same attitudes and emotional reactions to his/her work stresses, but he/she may have more physical and perhaps emotional stamina to deal with the inner effects of these reactions. For example, an IT worker may experience disappointment, frustration, and insecurity as a result of not being able to meet deadlines and satisfy user expectations. After changing his diet, working out regularly, and intentionally taking time for recreational family activity, this worker will still likely experience disappointment, frustration, and insecurity in the same work circumstances, only now the effects of these experiences on the person will be lessened.

The question is: What can be done so that this IT worker will no longer react in disappointment, frustration, or insecurity in the same work circumstances? A change in this worker's *inner reactions* to the same work environment must be brought about. Such a change in consciousness can be achieved through increased psychological awareness or "enhanced emotional intelligence."

Firstly, it is worthwhile to consider more specifically the effect of Myers-Briggs personality consciousness on worker stress in IT. An earlier mentioned incident is here worth re-examining. Two IS employees from a local finance company came out as ISTJ (introverted, sensing, thinking, judging {structured}) and ISTP (introverted, sensing, thinking, perceiving {open-ended}) types respectively. When I asked each, "What situations in your work stress you the most?", the ISTJ responded that she was very stressed when she had "too many things on her desk at once." She preferred to work on one thing and finish it (Js like the closure) and then move on to the next. The second person, an ISTP, said she was very stressed when she had only one thing to work on—she needed to be involved in several things simultaneously. The adage "one man's food is another man's poison" was truly exemplified here.

System development comprises a variety of tasks, and it is indeed likely that some personality types will be stronger in certain tasks. An MBTI consciousness in the IT organization could bring such truths to light. People would be able to express openly to co-workers and supervisors why specific tasks are easy or difficult for them and where they would appreciate the assistance of another specific, complementary type. IT workers could also monitor their energy level, noting which aspect of their work seem to drain energy, and they could then relate this drainage specifically to MBTI considerations. Those in IT management could become more explicitly aware of how different types are energized and stressed. This awareness could become a regular part of their management skills and strategies.

In addition to Myers-Briggs personality, cognitive style can have an influence on an IT worker's stress. It is easy to see that people of both analytic and intuitive (heuristic) cognitive styles are valuable in the IT profession. However, if people are not specifically aware of their styles and those of co-workers, needless stress can result. Organizations may indeed suffer productivity losses in system development due to task and cognitive style mismatch. It is hoped that research will address this question directly in the near future. Also, an employee's professional self-confidence may be eroded when, for example, a promising intuitive is mis-assigned, and then poorly evaluated by a highly analytic manager.

Since stress often results not so much from the external event itself, but from the internal emotions it evokes, the Enneagram personality system, with its nine types and nine underlying emotions, can be of use. It would not be far-fetched to assume that the IS field contains many Ones (Perfectionists), Fives (Knowledge Seekers), and Sixes (Loyalists). A Perfectionist will have difficulty being "perfect" with not enough time to do so, yet if this need is thwarted, his entire psychological existence may be at stake to him. A Knowledge Seeker can become overwhelmed with having to learn too quickly without adequate time for deep comprehension, yet his main fear, which can become "existence threatening," is that of being overwhelmed by the environment. A Loyalist wants to do what the "externals" dictate, yet she may face unrealistic deadlines and expectations. With a central need to be loyal to their expectations, this type of IS worker may indeed be headed for burnout unless she is explicitly conscious of her limiting attitudes. She then needs clear prescriptions, such as the Enneagram's "directions of integration," for transcending these attitudes.

While growth in awareness of personality traits and cognitive style can provide definite resources for dealing with common IS stresses, the conscious awareness of the deepest inner self, if properly developed, can provide a quantum leap in an IS worker's psychological robustness. However, this type of consciousness, for many, may indeed be most difficult to develop. It requires an actual shift in awareness from the intellect-emotion-body level to an entirely deeper level. Moreover, such a shift cannot be achieved solely by willpower, insight, or a changed way of thinking. As well, one cannot intuit what it would be like to experience one's deepest self until one is actually in the experience. According to Dr. Bernie Siegel, this "perfect self" is not located in the "conscious mind"—it is real, yet trans-intellectual. It is likely that significant awareness of such a level has been eclipsed, at least to a fair degree, by an almost exclusively intellectual focus following the Industrial Revolution in the Western world.

The key factor in such inner growth is a shift in *awareness*, by an opening of the "inner eyes." As mentioned earlier, it is this awareness that gives intelligence to emotion. By analogy, it is like seeing a movie in color on TV that until now one has seen only in black and white. The scenes are still the same, but the experienced content is much richer. One can, for example, pass by a flowery bush every day; one can even enjoy the sight, on an emotional level. But one day, upon a deep inner opening, one will experience a new, much richer awareness of the beauty and life energy emanating from this flower—one may then simultaneously experience a convincing consciousness that the beauty, dignity, harmony, and life energy in that flower is the same as the beauty, dignity, harmony, peace, and deeply joyful life energy at the level of one's own deepest inner self. *Such an experience is indeed transformational!* One has now seen for oneself, with one's own inner eyes, what one has until now seen mostly (or only) through the outer eyes.

In other personal growth approaches, one usually needs to do specific work to *develop* something—a new way of thinking, a new set of values, a new capability to "switch" to another mental framework. In growth in the deepest self, there is nothing to develop, everything is already present gratuitously, deep within. The growth work lies in *discovering* this revitalizing reality, which gives tone to thoughts, feelings, and actions—unblocking, analyzing, then deeply feeling and bodily sensing those experiences (or lack thereof) in one's life history, which may have largely eclipsed one's consciousness from one's riches and most stabilizing inner resource.

It should be pointed out that, at the start, looking deeply within may appear "scary." This may be because one is connecting with blockages (the wedges in Figure 4) rather than the deepest self. The experience of the deepest self is far from "scary"; in its advanced form, it surpasses, in depth and richness, by far, any exhilaration one would get from suddenly winning a $15 million lottery, being named to a very prestigious position, or (for single persons) becoming engaged to the most attractive, wealthiest, most humorous, sensitive, and highly intellectual person in the world. Yes, such indeed is the full power of the deep inner core—it is authentic, revitalizing power; it is "the indestructible, diamond-like, positive, loving, spiritual essence" that Bob Hoffman invites people to uncover. A person living and working with this type of inner connectedness, even only to a "fair" degree, will naturally be authentic, attractive and inviting, and far from "spaced-out."

How, specifically, can connectedness to the deepest self help with IS stress? In relating emotionally to one's work, a deeply connected person has the capacity to be both *close* and *free*. One can be intimately involved in developing a data model, testing a program or dealing with an unexpected disruption, and after leaving work, one can peacefully "drop oneself into" the revitalizing energy of one's inner core, freeing oneself from lingering negative emotions and their physiological consequences. There will still be stresses, frustrations, and disappointments at the emotional level—these may indeed hurt—but such pain will never eradicate the deep inner security and life energy (the transmission may be in trouble, but the engine will be solid).

It may indeed be construed that burnout results from reactions that give rise to lingering emotions, which block inner access to the core of one's *per-*

ceived existence. If, for example, one perceives the heart of himself to be a competent project manager and derives a good part of his life energy (raison d'être) from this perception, a missed major deadline can indeed initiate emotional eclipse from his major energy source. If such emotionally challenging experiences occur repeatedly, one is repeatedly "gasping for emotional air" and coming up short. Growth-in-being shifts the major inner energy source from a *perceived* locus (such as consciousness of one's competence) to an innermost *real* source which no external event can ever extinguish. Hence the potential for inner robustness and resiliency.

As well, a deep inner connection provides inner strength to act with unquestionable *authority*. Since fear is present at the level of the emotions, but never at the deepest inner level, an integrated IS manager will not be hesitant in expressing unmistakably, firmly, yet cordially that this request can *not* be met, that this workload can *not* be taken on, no matter what pressure comes from outside. Moreover, such a statement, made with truthful authority, is usually heard, respected, and accepted. An IS manager may well *react* from her emotions, but she will *respond* in truth and firmness from her deepest inner core. Thus, an important question for an IS worker growing in deep inner awareness who is met with work stress is: "*From which part (mostly) within me am I experiencing this situation?*"

As mentioned earlier, a systematic, effective, well-established program to guide persons into their deepest inner self is the Hoffman Quadrinity Process, offered worldwide by the Hoffman Institute. Michael Ray, Professor Emeritus of Creativity, Innovation, and Marketing at the Stanford University Graduate School of Business, says:

*"Everything changed after the Hoffman Quadrinity Process, because I then understood the obstacles to my full potential…It presents a powerful combination of practical approaches that have been honed to a package of **incredible power that stays with you**…Any accomplishments that someone might list after my name came largely because of this experience" [bold added].*

It would indeed be fitting for a number of adventurous IT executives who would like to explore this dimension of inner growth and stress management to be sponsored for the Hoffman Quadrinity Process. There is no doubt many such executives, after two years or so, will have become significantly empowered from a totally new inner dimension. Then, testimonies of practical IT work-related results could indeed initiate more widespread interest and eventually acceptability of deep inner growth within the IT profession.

To summarize, stress in IT is not likely to disappear in the near future. However, through emotional intelligence enhanced by psychological awareness of personality and thinking style, one can lessen significantly the negative effects of stress. Once one has stared in the face of the monster in truth and inner connectedness, the monster loses its power.

HUMAN RESOURCE ISSUES

Apart from influencing IT work, awareness of psychological dynamics can also influence the management of IT workers. Issues such as hiring, retention, mentoring, promoting, and gender balance can be addressed with enhanced emotional intelligence.

In Chapter I, the issue of using Myers-Briggs typing in *hiring* for IS professionals was discussed. While it can be questionable whether to require an applicant to complete an MBTI questionnaire, an astute interviewer can ask work-related questions that may indeed bring out preferences for Introversion/Extraversion, Intuition/Sensing, Thinking/Feeling, and Judging/Perceiving. Such unconfirmed perceptions can nonetheless form part of the "whole picture" on which the applicant is assessed. Another area to note, on the part of an interviewer, is emotional motivators. From this, one can anticipate what work situations might be most distressing for the applicant. Degree of emotional dependency on one's job and the source of one's main identity may indicate potential for resiliency. Even interest in psychological factors and "soft skills" may foretell the candidate's future ability to develop as a "whole person."

Initial impressions regarding psychological dimensions as suggested above can then be compared over a longer period of the hired person's employment. In time, HR persons with significant perceptive ability can prepare manuals for other persons in similar positions, explaining how they assessed an IS candidate's psychological profile.

An area that can really benefit from psychological awareness is *mentoring*. Firstly, the mentor would need to be psychologically aware, along the lines of Part I and beyond. His aim would then be to help his understudy to "grow his whole self," in the words of XP founder Kent Beck. The mentor could provide insight, suggestions, but above all *presence*. Addressing an understudy from one's deepest, most authentic, empowered self would begin to open, gradually, a deeper awareness in him/her. Organizations can develop a "mentor training program" when psychological factors such as those presented here are explicitly addressed. Effective mentoring with enhanced emotional intelligence may indeed be a main ingredient in the retention of valuable employees.

Further addressing *retention*, Dr. Linda Berens, founder of the Temperament Research Institute, maintains:

"Individuals seek satisfaction for a core need: i) to better understand the meaning and significance of one's life, ii) to seek mastery and to become universally knowledgeable and competent in whatever one undertakes, iii) to seek membership or belonging to a group and to fulfill responsibilities and duties for the group, and iv) to have the freedom to act according to the needs of the moment so as to make a unique impact on others or the situation."

It may be easy to see that the four needs correspond to the NF, NT, SJ, and SP Keirsey temperaments derived from MBTI. As well, the less connected a person is to the deepest core energy, the more acute such needs are likely to be. Thus, it may be wise for IT human resource persons to research different IT jobs so as to determine which needs seem to be best satisfied in which job. This can have definite impact on hiring, retention, mentoring, and promotion.

A wise IT organization will recognize that employees with different psychological characteristics will prefer to progress through different *career paths*. Explicit consideration of factors as outlined in Part I may motivate the enabling of a variety of ways one can progress in the organization, considering one's unique psychological makeup. Worthwhile research in this area would be to see if there is a relationship between prevailing emotions experienced and different IT jobs. Would it appear, for example, that a "typical" help-desk person, a trainer, and a programmer each experience some specific emotion in the course of their daily work. Thus, in career path planning, one can be shown that by moving from job A to job B, one would not only gain the following specific technical skills, but one would also develop the following specific dimension of one's emotional intelligence.

Apart from providing an employee with a "fitting" career path, an organization should attend to an employee's need to be respected and understood. Inwardly centered IT management in touch with one's own inner dignity and empowering energy will automatically be able to address subordinates from that framework. The deeper and more integrated a manager's psychological awareness, the more desirable he will be to work for. The goal, then, is to cultivate an emotionally intelligent work environment.

Another issue of concern is that of *gender equality*. Here the formulation of policies as well as their application, if carried out from the deep, empowering inner consciousness, can transcend prejudices, establishing fairness as the norm. On the other side, inward rootedness among members of the under-represented gender will give their efforts at establishing equality the true power of (inner) authority rather than the limited power of willed assertion or aggressiveness.

Deeper inner growth on the part of IT management can play a significant role in the enabling of *work/life balance* for IT employees. An inwardly connected person lives (to an appreciable degree) from personal wholeness. He is aware of his essential identity as a human being and does not see himself as a "job label." Such an IT manager will have a natural motivation for life balance as well as an inner capacity to achieve it. He will no doubt promote the same perspective for his subordinates.

Psychological awareness can indeed affect many areas of IT work, including those discussed in this chapter. The IT profession has entered the "post-modern era." Increasingly, it is being recognized that IT, to maintain impact and influence, requires a truly human, not just intellectual resource.

IT EDUCATION

Since it has been pointed out so far in this chapter that psychological factors, such as those introduced in Part I, can indeed contribute significantly in many aspects of IT work, it is appropriate to address IT educators with the hope of having them introduce this topic within their curricula. Psychological factors can be introduced, in varying degrees of thoroughness, in community college IS analyst diploma programs, undergraduate MIS and MBA-IS major programs at universities, as well as in a variety of Continuing Education settings.

A suggested starting point would be to have the class take the Keirsey Temperament Sorter, available online, if not the entire Myers-Briggs test. Students could then be segregated into the four temperament groups and asked to solve a larger IT problem. Each group could then present to the entire class the steps it took in arriving at its answer. Specific implications of type theory could be pointed out in this context.

If time permits, other factors could also be introduced, for example Enneagram types and cognitive styles. If this is a Continuing Education class for people who are regularly working in IT, students can be asked to do some concrete observation in their work environments. Then they may prepare appropriate presentations. Regarding material on the deepest self, basic ideas can be presented in a classroom format. Students can be familiarized with the orientation of programs such as PRH and Hoffman's Quadrinity, as well as with current personal development authors such as Dyer, Borysenko, McGraw, Siegel, and Zukav/Francis. Furthermore, it would be beneficial for students to relate the reality of the deepest self to their own life histories, perhaps in a "workshop" format. It may be difficult to have students actually develop deeper, lasting, experiential awareness along this dimension in a one-term course. However, handouts with "tips" for ongoing development could be provided at course end. In another setting, e.g., a four-year evening course program, material on psychological awareness could be provided in the first course of the first year. Students could be asked to maintain a "journal" of relevant work situations on this topic throughout the four years. They could then report on their discoveries in the final, capstone course.

It would also not be difficult to construct a one-term course specifically on psychological factors in IT, possibly using material here presented as a major reference. Such a course could be located within Continuing Education or graduate programs in IT management. As for specific content, the course could devote: three classes (perhaps three hours each) to Myers-Briggs and applications (as in Chapter I); one or two classes on the Enneagram and applications to IT (Chapter II); two classes on cognitive styles, creativity, and learning (Chapter III); two or three classes on the deepest self (Chapter IV); and then one or two classes on areas of application (Part II). Students would select one of the major topics on which to do a paper and possibly a presentation. As well, shorter, but more intensive approaches, for example a week-long course, could be developed along these lines.

The topics of psychological awareness and emotional intelligence, as specifically applied to various facets of IT work, do deserve to be recognized, at this time in IT history, as material worthy of serious consideration within many IT education frameworks.

CONCLUSION

This chapter has identified and addressed common aspects of IT work where increased psychological awareness could make an impact. It has indicated where and generally how the factors from Part I could be used to provide more effective IT functioning. Yet, readers may feel that more explicit direction would be in order. For example, identifying specific situations in requirements determination where types with different orders of preference would perceive specifically different system aspects could be very educational. Identifying exactly *how* two different MBTI types complement each other in an XP programming context could provide convincing relevance.

However, more specific empirical research is necessary to answer such questions with at least minimal authority. It is hoped that this initial effort, by providing a basic yet relatively thorough introduction to some influential factors and by proposing possibilities for their application, will motivate many concerted initiatives that would give rise to more detailed, specific, and impactful presentations.

REFERENCES

Alavi, M., & Joachimstaler, E.A. (1992). Revisiting DSS implementation research: A meta-analysis of the literature and suggestions for researchers. *MIS Quarterly, 16*(1), 95-116.

Barki, H., & Hartwick, J. (1994). User participation, conflict, and conflict resolution: The mediating roles of influence. *Information Systems Research*, 5(4), 422-438.

Berens, L., & Isachsen, O. (1995). *Working together.* Institute for Management Development.

Boehm, B. (1981). *Software engineering economics.* Englewood Cliffs, NJ: Prentice-Hall.

Brod, C. (1984). *Technostress.* Addison-Wesley.

Fitz-Enz, J. (1978). Who is the DP professional? *Datamation, 24*(9), 125-128.

Glass, R. (1997). The ups and downs of programmer stress. *Communications of the ACM, 40*(4), 17-19.

Green, G. (1997). Examining the impact of cognitive complexity on project integration performance of project leaders. *Proceedings of the Americas Conference.*

Henderson, J., & Lee, S. (1992). Managing IS design teams: A control theories perspective. *Management Science, 38*(6), 757-777.

Kaiser, K., & Sirivasan, A. (1982). User-analyst differences: An empirical investigation of attitudes related to systems development. *Academy of Management Journal, 25*(3), 630-646.

Katzenbach, J.R., & Smith, D.K. (1999). *The wisdom of teams.* Harper Business.

Khosrowpour, M., & Culpan, R. (1989-90). The impact of management support and education: Easing the causality between change and stress in computing environments. *Journal of Educational Technology Systems*, *18*(1).

Kolb, D.M., & Sheppard, B.H. (1985). Do managers mediate or even arbitrate? *Negotiation Journal, 1,* 379-388.

Larsen, D. (2003). Embracing change: A retrospective. *Cutter IT Journal, 16*(2), 39-46

McCarthy, J. (1995). *Dynamics of software development.* Microsoft Press.

Newman, M., & Robey, D. (1992). A social process model of user-analyst relationships. *MIS Quarterly*, (June), 249-266.

Riley, M. (1990). *Corporate healing.* Deerfield Beach: Health Communications.

Robey, D., & Franz, C.R. (1989). Group process and conflict in system development. *Management Science, 35*(10), 1172-1189.

Stokes, S. (1996). Coping with change at the top. *Information Systems Management*, (Winter).

Wastell, D., & Newman, M. (1993). The behavioral dynamics of information system development: A stress perspective. *Accounting, Management and Information Technology, 3*(2).

Chapter VI

The Emotionally Intelligent IT Organization

INTRODUCTION

At this point it may be useful to hypothesize how a typical North American IT organization might evolve in psychological awareness/emotional intelligence. Such a vision may be useful in showing IT managers, in encapsulated form, what may indeed be possible.

In 1974, Richard Nolan identified six stages of data processing growth within an organization: initiation, contagion, control, integration, data administration, and maturity. Here, a similar attempt is made to identify and describe *growth stages in enhanced emotional intelligence* within an IT organization.

A GROWTH STAGE MODEL

Initially, before the first stage, no formal attempt has been made to introduce psychological factors to the IT organization. A few individuals may have read some books on personality or motivation and may have discussed some of this material privately with receptive co-workers.

Stage 1: Orientation

Members of the IT department are given a Myers-Briggs personality type seminar. Each person has his/her type assessed. The presenter speaks generally of organizational applications but does not address IT specifically.

In the following weeks, some IT staff discuss their types with co-workers, a few "pioneers" attempt to address basic issues (e.g., extraversion, intuition) regarding teamwork. Many discussions are, however, light-hearted.

Articles on MBTI relevance begin appearing (perhaps twice a year) in professional IT journals and newspapers. Such articles may be posted on the coffee room bulletin boards by astute IT managers.

After a while, the occasional project manager (likely an NF temperament) will, mostly informally, consider personality preferences when assigning people to tasks. Some IT managers, on their own initiative, may purchase a book such as this one and consider at least half-seriously the possibilities presented.

This is the stage of orientation. People begin to become aware of their preferences on the four MBTI dimensions. The Myers-Briggs system, because of its structure and adaptability in wide circles, is likely the best starting point for psychological awareness in IT.

To move an organization beyond this stage, more relevant articles appearing frequently in professional literature will definitely help. Applied research and development (from "Psycho-informatics" researchers collaborating with senior MBTI trainers) may yield literature applying MBTI directly to IT. For example, a booklet parallel to the existing one on *type in organizations* can be produced on *type in IT organizations*, outlining main work characteristics, strengths, weaknesses, and areas for development for each of the 16 types. Such an effort would indeed be very valuable to get organizations to progress beyond the first stage.

Stage 2: Consideration

After reading recurring articles on the value of MBTI in IT, after attending MBTI workshops at professional conferences and reading newly developed IT-specific Myers-Briggs material, IT management decides to "get serious."

An IT-specific Myers-Briggs workshop is given. Management circulars encourage personality type consideration in IT work. Persons on a team begin to discuss openly personality issues as related to cooperative efforts.

In some organizations, specific applications are described in an internal newsletter. Particularly, applications that imply increase in effectiveness begin to be noticed by management.

By this time, research and development efforts (perhaps initiated by the formation of an "Institute of Psycho-Informatics," composed of MIS academics and Myers-Briggs trainers/psychologists) will have produced a booklet and workshop on personality type and user training, focusing on users' learning styles and the development of training efforts to accommodate different styles. Personality/learning styles are considered in user training. Successful efforts are publicized.

Until now, nearly all psychological considerations have revolved around the Myers-Briggs system. Other psychological dimensions are discussed here and there, at coffee breaks.

The main characteristic of this stage is the serious acceptance of at least one psychological dimension, Myers-Briggs personality, as having a specific relevance to IT work. To achieve this level of involvement, widespread communication of specific, impactful efforts is required. As well, a select group of Myers-Briggs trainers, e.g., from the Association for Psychological Type, can recognize significant, concentrated potential for MBTI application, and begin to develop workshops and *materials oriented specifically to the IT field*. In time, a few such people, coupled with several applied academics, may indeed form an "Institute of Psycho-Informatics," which could have a research arm and a development arm, and could be dedicated to the systematic introduction of psychological factors leading to enhanced emotional intelligence in the vast IT industry.

Organizationally, to move from the first to this second stage, it will be indeed useful to have a "champion," a person in higher middle management (quite possibly an intuitive feeler). He/she would be willing to publicize existing application in the industry and to promote, specifically, a more concentrated involvement within his/her IT area.

Stage 3: Initial Acceptance

By this time, IT people have been using MBTI, at least semi-formally, in task assignments, team cooperation. and user training. Now the use of MBTI expands to system requirements determination with users and to career-path planning.

Development of specific, visual materials (e.g., on a CD), to train analysts how to recognize communication trends among different types of users, can be very motivating and very useful.

At this point, Human Resources personnel dealing with IT become more strongly involved. Policies on using MBTI formally (e.g., in interviewing) are established. Also, *cognitive style consideration* becomes more widespread. Kirton's KAI test may be formally administered to IT personnel. Considerations related to creativity may be initiated.

Myers-Briggs personality and cognitive/creativity styles are applied to specific IT methodologies, such as Extreme Programming. At this stage, IT personnel are beginning to feel "at home" with such psychological dimensions. They begin to accept them as part of their job consciousness. More time is devoted to such psychological topics at professional conferences, and more print space is devoted in professional publications. Organizationally, a specific "Psych Factors" electronic newsletter may be produced and distributed quarterly. In the industry, a specific IT Psych Factors website is established. Specific comments from individual IT workers on how such consciousness has made their work become more effective (and, hopefully, more enjoyable) become most noteworthy.

While Myers-Briggs personality, cognitive style, and creativity modes will have become, at least to a fair degree, acknowledged and accepted in the IT organization, issues regarding emotions will not yet receive formal attention. However, the communication sources to which IT workers will now be looking acceptingly for further development on MB and cognitive awareness may now begin to expand their content into areas such as Enneagram types and emotional intelligence.

Some IT organizations may, thus, be ready to add, for example the Enneagram to their psychological repertoire. The Enneagram is particularly useful in moving a structured, analytic person into initial emotional considerations. Firstly, it is a personality typing system, and having acknowledged Myers-Briggs, another system will not be intimidating. Secondly, it is structured (and numeric), with its own diagram. Thirdly, it postulates an *underlying emotion* for each type as well as an emotional pattern. Here is the point of greatest progress for an IT worker in this third stage. Persons will begin to see *how their emotional patterns have both enhanced and hampered their work.* Fourthly, the Enneagram contains specific, prescribed growth paths for each type, implying the capacity of each type for emotional development.

At this point, IT workers are beginning to recognize and discuss previously hidden emotional needs that have largely motivated their work. They begin to notice dysfunctional emotional reactions while under stress, and relate this to established videos and Enneagram-in-IT guides which will have been developed, possibly by the IPI.

The topic of emotional literacy begins to surface increasingly at professional conferences and in trade publications. More IT staff take note of significant improvements in effectiveness due to awareness of Myers-Briggs, cognitive style, and Enneagram patterns. Relationships of such awareness to stress management is noted.

Some individuals (again, likely intuitive feelers) have read Goleman's *Emotional Intelligence at Work* and are beginning to discuss possible applications. At many IT organizations, a "Psych Factors Group" (PFG) meets, voluntarily, every three months; participation is increasing.

This stage is characterized by a more formal entrenchment of Myers-Briggs as a "soft skill" in IT. Cognitive styles are attended to. More significantly, though, an explicit open focus on underlying emotions and emotional response patterns is made possible by the Enneagram. Such a direct focus on emotions is a great step forward for the heretofore exclusively analytic IT workers. It will open new doors in the next stage.

The key requirement in this stage is recognition, in specific instances, of the noteworthy impact of Myers-Briggs awareness on various aspects of IT work. IT workers will wonder: If one type of psychological awareness can indeed make a significantly positive impact, what about others?

Stage 4: Emotional Focus

At this point, Enneagram applications receive more notice. It becomes "in vogue" to address emotional considerations in IT work (NF types are rejoicing—they are finally getting their due). The psychological awareness movement within the IT organization has taken hold.

Considerable focus is now placed, at least semi-formally, on *emotional intelligence*. IT workers are becoming more aware of their own feelings and of those with whom they work. Internal IT newsletters refer to specific applications. An industry-wide website provides literature, Q&A boards, and discussion groups.

Now, a variety of other psychological techniques are examined: neuro-linguistic programming, focusing, and meditation. These approaches are related to stress management.

Some Psych Factor Groups meet more often, with more fruitful discussion. The Institute of Psycho-Informatics will have produced numerous materials related directly to IT; its staff will be providing a variety of workshops. As well, higher management has taken note of this undeniable movement to "soften and broaden" IT. In some organizations, higher IT managers are sponsored for the Hoffman Quadrinity Process, in an attempt to "go all out."

In this stage, the IT organization has, as a whole, said "yes" to psychological awareness/emotional intelligence. **Calls for "growing the whole person" (as made by Kent Beck) and for "empathy…that provides visceral motivation," as promoted by Jim McCarthy, are seen as central to IT's effectiveness within the organization.**

The role of NF types (intuitive feelers) in IT's arriving at this stage cannot be underestimated. Such psychological awareness is the *natural preoccupation* of this type. They will truly have the opportunity to "come into their own" at work. However, NFs will need to learn, based on applied research, how such awareness can be developed and promoted to types other than their own. This can be a major challenge.

Stage 5: Intended Wholeness

At the start of this stage, the organization is actively "working on" emotional intelligence. However, the effort is largely made from an intellect-emotion-body perspective on the human person.

Still, a number of higher IT managers, who have completed the Hoffman Quadrinity process, have now realized another, deeper dimension. Such IT Hoffman graduates form their own "fraternity." In time, the perspective of the *deepest inner self* is promoted more directly.

More IT employees are sponsored for the Hoffman process and similar being-centered growth efforts. Deep authenticity of IT workers is promoted, in policy and in fact. *"Growing the whole person" now becomes an IT strategy*. Psych-Factor Groups now become Inner Growth Groups, where not only valuable *ideas* are shared, but *feelings* are expressed and deeper inner energies are reached. The multi-dimensional welfare of the IT worker becomes of paramount importance: stress issues are addressed head-on.

Retention of employees is increased as they feel respected and addressed as "whole persons." Conflicts are handled more fruitfully. Creativity is increasing, as is motivation and morale. IT professionals are forming an integral part of international "Spirit at Work" movements.

Many employees at this stage have appropriated, to a fair degree, the psychological factors promoted in this book, as well as valuable others. *The days of the "techie-nerd" are long gone!* The IT organization at this point is wholistic and emotionally intelligent.

INTEGRATION WITH PEOPLE - CMM

The Software Engineering Institute (SEI) had promoted the SEI-Capability Maturity Model (SEI-CMM) as a growth stage model for evolving the process of software development,. Then, in order to address *people* management among software developers, it came up with a derivative, the People-Capability Maturity Model (P-CMM). This is an organized plan for progressing through evolutionary stages of *managing the people who develop the software*. It can also be applied to other work settings.

P-CMM's primary objective is to improve the capability of the workforce. It is based on the best current practices in fields such as human resources, knowledge management, and organizational development. It has been used worldwide since 1995 by large and small organizations (including IBM, Boeing, Lockheed Martin, and Ericsson). The People-CMM helps organizations *characterize the maturity of their workforce practices*, establish a program of continuous workforce development, set priorities for improvement actions, integrate workforce development with process improvement, and establish a culture of excellence.

P-CMM identifies five maturity levels: initial, managed, defined, predictable, and optimizing. Also there are four areas of concern that grow through these five levels: developing individual capability, building work groups and culture, motivating and managing performance, and shaping the workforce (e.g., staffing, workforce planning).

Thus it appears that there is definite potential for *incorporating development of psychological awareness into the P-CMM model*, so that the intended "culture of excellence" would include emotional intelligence.

The model refers to specific workforce management elements such as mentoring, empowering workgroups, career development, work group competency integration, and continuous capability improvement for workers and groups. Each of these areas could make reference to incorporation of psychological awareness.

Indeed, such an enhancement of P-CMM may occur in time. This would truly be a sign of the IT industry's acceptance of psychological awareness as a dimension of the maturity of the IT workforce.

CONCLUSION

The psychological emancipation of the IT profession, as hypothesized in this chapter, will not happen overnight. The identified five stages above could well take 15 years to implement. However, the stage appears set for starting. Myers-Briggs particularly, and the Enneagram to some degree, have been considerably accepted in professional and personal life. "Self-help" books are proliferating. Spirituality (authenticity) at work is gaining respectable attention. To quote again consultant Martin Rutte (who has contributed to "Chicken Soup" books, although he has not yet written *Chicken Soup for the IT Professional*):

"We're in a paradigm shift...there will emerge...new ways of work. Environmental degradation and lack of fulfillment are coming to an end. Respect and calling forth of people's individual gifts—spirituality—that's what's coming in."

With the threat of burnout epidemic, the IT profession may indeed be facing a psychological ultimatum: *breakdown* or *breakthrough* to a new vision, new awareness, and new experience.

REFERENCES

Curtis, B., Hefley, W.E., & Miller, S.A. (2001). *People Capability Maturity Model (P-CMM) Version 2.* Software Engineering Institute, CMU/SEI-01-MM01.

Nolan, R.L. (1982). *Managing the Data Resource Function.* West Publishing.

Chapter VII

A Call to Action

INTRODUCTION

This book was not written primarily for academic interest (or for entertainment). Given existing interest among IT workers in some psychological factors and calls from respected professionals for further development, *the book is intended primarily as a catalyst to further action* regarding comprehensive psychological awareness in IT. To this end, specific suggestions for action are now outlined for individual IT workers, employing IT organizations, university academics, and professional IT associations. Possibilities for collaboration on innovative efforts are also presented.

INDIVIDUAL WORKERS

One thing that an individual IT worker who has even a slight interest in psychological awareness can do is to read this book. If he/she has indeed found enough valuable insight or food for thought, for feelings and for questioning, the person can recommend it to co-workers/supervisors. It is hoped that a specific website will be established to field questions and accept suggestions relating to this book.

As well, the interested individual can examine reference material that may seem appealing or intriguing. After possibly having zeroed-in on a topic that may have significant potential for application at work, the person can "experiment" and observe. After some time, he/she may have indeed discovered a new awareness or attitude that is proving helpful at work. Depending on the potential for receptivity in one's work culture, one may share this insight with co-workers, investigating the potential for further, collaborative observation. Eventually, if enough persons take an interest, specific action (e.g., a Myers-Briggs workshop) can be suggested to IT management.

EMPLOYING IT ORGANIZATIONS

Having considered the factors in this book, senior IT managers must assess the "operational feasibility" of promoting, at least to some degree, further psychological awareness in their current IT culture. Admittedly, in some environments, this will not be the time, for various reasons. If, however, some "exploratory" involvement seems possible in the near future, the senior managers might bring up the topic with middle managers (at least those who may be initially receptive), more or less formally.

After having considered the proposed five stages in Chapter VI, a group of IT managers may wish to establish a preliminary plan as to what they would like to do in this regard for the next one to three years. They may wish to "poll" informally influential IT workers to gauge potential response to their planned efforts. Thus, using the five stages as a base (or perhaps changing the order suggested therein, if they feel it is appropriate), they can plan more specific activities (workshops, discussion time, electronic suggestion boxes, etc.). A desired result could accompany each planned activity.

Significant factors influencing the choice to proceed at least in some way will be workload and stress level among IT staff. Initially, the higher the workload and stress level, the less inclination IT management will have to get involved in psychological development. At times, however, the converse may be more appropriate. If one is noticing impending burnout among several key IT staff, a radical approach to their "personal re-engineering" may be in order. The staff need to be given a longer time to breakthrough before they completely breakdown.

To this end, a concerted effort may be initiated to provide, for example, at a specific "retreat center" a program for one- to three-month "personal re-engineering sabbaticals." IT staff nearing a breakdown would be sent to such a place that would specifically cater to stressed IT workers with wholistic approaches for physical, emotional. and spiritual rejuvenation. Here, personality type and emotional patterns could be specifically addressed, along with guided programs for deeper inner connectedness. Such staff, upon return from their "sabbatical," may be the catalysts for moderate yet definite organizational change in IT, based on psychological awareness.

With any efforts introduced for the IT staff, management will need to monitor the perceived effects, possibly comparing them with similar observations by managers from other organizations. Such observations could be shared even anonymously at a dedicated website.

PROFESSIONAL ASSOCIATIONS

Professional associations of IT workers, such as for example the Canadian Information Processing Society, can also play a role in the movement towards psychological awareness.

At least one (or two) monthly meetings per year can have a specific psychological topic as its theme. Guest speakers, presenting results from successful application, can indeed motivate significant interest. Annual conferences can have a track related to psychological awareness, consisting of presentations, panels, and workshops. Newsletters can have a regular "Psych-IT" column. Competitions could be held for submitting a written case outlining a successful psychological application in IT.

The associations could also help to fund development of relevant training materials (e.g., workbooks and CDs) on specific aspects of psychological awareness. For example, a video-CD can be developed outlining, through interviews with specific people, how to recognize user thinking style when eliciting system requirements (on the video, requirements for the same system could be articulated by persons with four different styles).

Professional societies could also contribute to committees planning specific stress assistance for IT workers, such as "personal re-engineering sabbaticals" mentioned earlier. Such societies could also network with counterparts around the world in addressing specific issues (e.g., work/life balance) through psychological development.

In short, IT professionals' societies need not be passive observers in the psychological emancipation process within their profession.

UNIVERSITY ACADEMICS

Practical application of psychological factors in IT would be very haphazard without the solid foundation of academic research. In the area of psychological factors, some significant research efforts have been noted (among them those referenced in this book). However, studies have tended to address certain dimensions more than others.

For academics to expand their research into various dimensions as outlined in Part I, they may need to see significant interest within the IT profession. As well, research paradigms, methodologies, and constructs are more established for cognitive style studies than they are for Enneagram growth or influences of the deepest inner self. Future research in application of psychological awareness to IT work will need to be interdisciplinary, likely involving information systems, psychology, and eventually the emerging academic field of human spirituality. Some departments of computing/IS do have a psychology PhD on their staff (e.g., University of Manchester), but such a person would likely be involved mostly in cognition studies, artificial intelligence, and human/computer interaction. If the involvement in psychological factors in the IT profession does proliferate, a particular research discipline, tentatively called here "Psycho-informatics," may eventually be established.

IS academics working to research psychological factors will need to team up with industry, such a collaboration likely yielding very interesting and fruitful results. At IS conferences, more attention can be focused on psychological applications of personality and emotional intelligence. Interested researchers can also make impactful presentations at applied spirituality conferences such as the Spirituality, Ethics and Work Conference, which has been held at the HEC in Montreal.

In terms of teaching, MIS faculty can begin to include enhanced emotional intelligence factors in courses on IS management. Teaching cases involving the topic can also be developed, based on industry experience. As well, senior undergraduate and graduate MIS/applied computing students can be exposed to factors as in Part I, so that they enter the workforce with a more comprehensive orientation.

PARTICULAR INITIATIVES

While university academics are called upon to develop solid conceptual foundations for psychological application in IT, this area can benefit significantly from *action research*. Action research is essentially action and research on the action that is carried out by the same person in parallel. More formally stated, "action research is a research method the essence of which is the juxtaposition of action and research, in practice and theory, through cyclic execution of four characteristic phases: planning, action, observation, reflection" (Estay-Niculcar & Pastor-Collado, 2000). AR has been considered a distinctive form of research since the 1940s. Recently its applicability to the area of Information Systems has begun to be recognized.

There is considerable potential for action research where an individual worker or IT manager is applying a psychological factor such as Myers-Briggs type in an IT situation. The person can then "stand back," observe and reflect on "what is going on," and extract valuable knowledge from the process.

Another specific effort, as mentioned earlier, would be the formation of an "Institute of Psycho-Informatics," which would have a research arm and a development arm, the latter producing specific training materials and workshops to facilitate consideration of psychological factors in IT. Such an Institute can perhaps initially "train trainers," who could then deliver a uniform psychological awareness program across the continent.

A center for "personal re-engineering sabbaticals" may be established for IT professionals otherwise headed for burnout, perhaps at some fitting "retreat center" to provide a setting for re-focusing, possibly for one to three months, say during the summers. Such a setting could itself facilitate considerable action research on the part of interdisciplinary, wholistically oriented staff.

As more work is done, at various levels, the substance of such efforts will *need to be widely disseminated.* A comprehensive, centralized website concerned with overall application of psychological factors to IT is indispensable. Contained within it would be essays, survey questionnaires, references to appropriate resources, a bulletin board, Q&A section, as well as several chat rooms dealing with specific factors. Such a site can undoubtedly accelerate progression through the growth stages proposed in Chapter VI. Also, other books will need to be written, following up on this initial effort.

REFERENCE

Estay-Niculcar, C., & Pastor-Collado, J. (2000). The realm of action-research in information systems. *Proceedings of the BIT 2000 10th Annual Business and Information Technology Conference,* Manchester, UK, November 1-2.

Conclusion:
Part II

A book that introduces psychological factors to the IT world should certainly, at some point, address specific applications. Part II has aimed to do this, albeit in a "preliminary" way. Bearing in mind the number of IT professionals worldwide and the available options for telecommunication, it would not take long indeed for a number of noteworthy anecdotes "from the trenches" of IT efforts to surface. With this, there would be specific opportunities to test out the suppositions introduced in the last three chapters. Within a reasonably short time, more specific management guidelines could be developed regarding psychological factors' role in "Making IT Work."

The challenge will be to get through the first stage, orientation. Will the growth-prone IT workforce collectively accept at least the possibility that a deeper self-awareness along several psychological dimensions can significantly increase work effectiveness and decrease work stress? On the surface, such a hypothesis should seem at least plausible. But, is there a "critical realization" that the typical IT worker will need to make within him/herself in order to move towards at least serious consideration of such a hypothesis? Is there a critical "vision shift" required, and what might be the external and internal impediments to such a shift?

Undoubtedly, many IT workers are "thinkers" and they favor logic strongly. It may indeed be possible that a large number are living from the viewpoint of "I Think—Therefore I Am." But, if so, is this orientation a *conscious choice* or a *default option*? Are IT workers firstly logicians or firstly human beings? Have they, in general, stopped to consider whether they are masters or slaves to the requirements that their work places on their intellects? What effects has the IT work culture had on various dimensions of their human, psychological development? Might such questions initiate the "emancipation" of IT workers into wholistically mature human persons who can indeed be very close to, yet truly free of their work involvements? Is the time ripe to promote, in a concerted way, greater psychological self-awareness within the IT field? If so, *how* could such an ideal be promoted to reach the current IT mindset?

Answers to questions such as those above may indeed be central to the growth and potential influence of such a prominent profession throughout many parts of the "global village." What, in general, will comprise the psychological makeup of the "information resource facilitator" of the future, as compared to the "DP professional" of the past? It is indeed hoped that many possibility thinkers within IT will consider the visions outlined in this part, reacting in agreement or in challenge, but in either way crystallizing concrete steps towards a new era.

FINAL COMMENTS

While calls for "emotional literacy" and other forms of applying psychological awareness in IT work have been appearing more frequently in recent years, such calls have been rather sporadic. While specific, concerted efforts have been made to connect cognitive psychology to areas such as human/computer interaction, concentrated applications of personality psychology or of human spirituality to IT work have not been popularized. In this book I have attempted to "pull together" areas of existing involvement, to suggest yet new areas and to present an initial, at least somewhat coherent vision and framework for, essentially, extending and humanizing the sphere of IT work.

For some, the idea that involving "the whole person" in primarily intellectual work will make such work more effective may indeed be novel and possibly questionable. And, yet, we recall the challenge from an initiator of a programming paradigm to "learn and grow our whole selves"; a vision from a software "guru" of a project leader who can "resonate with the team's complex emotional state," resulting in a state of team empathy during the software development process; and a presentation from an IT consultant of the "emotional intelligence" that it took to have the inner strength to refuse an unreasonable request from higher management. We have also seen a quotation from an unquestionable IT authority on the prevalence and severity of burnout among software developers, where "some IT organizations have become almost completely dysfunctional."

In such a work climate, does IT not need complete "human" "beings"? Would it make sense to ignore aspects of the human person that could provide robustness and resiliency? It may be indeed noteworthy that, while the Industrial Revolution may have moved the human person into intellectual predominance, the "IT Revolution," with its recent calls for addressing and involving the "whole person," may indeed be initiating a re-centering of the human being in his/her essential core, giving rise to new consciousness, new vision, and new, empowering experiences. May this book encourage the first few steps along a novel and vivifying path!

Appendix:
An Integrative Model

INTRODUCTION

In light of the fact that several different psychological factors have been presented here for consideration by IT professionals as well as interdisciplinary researchers, it may be interesting and also useful to get an appreciation of a highly developed, conceptual "meta-model" approach to human decision making that has incorporated some of the above topics. It is all the more interesting as this model has been developed by a researcher in the MIS field, Dr. Cathal Brugha of Ireland, who has also been prominent in the European Management Science community.

The following is an excerpt from Dr. Brugha's extensive research on "nomological maps" and their wide applications, particularly in IS and decision sciences. Although this excerpt is itself quite brief, it does provide the reader with a "flavor" of the depth and sophistication of existing research on psychological issues that can be applied to IT.

A SYSTEMS OVERVIEW OF PERSONALITY-RELATED SYSTEMS

Cathal M. Brugha, PhD, MSc, MBA, Management Information Systems, Smurfit and Quinn Schools of Business University College Dublin Dublin, Ireland. Cathal.Brugha@ucd.ie

Nomology, the science of the laws of the mind (Hamilton, 1877, pp. 122-128), is a meta-model whereby issues such as management, personality, and spirituality can be considered. The basis of Nomology is that decision-makers tend to analyse problems which involve qualitative distinctions by breaking them into activities, or categories of behaviour, which are each important in themselves and follow natural sequences. This is a natural approach that the mind uses when addressing a problem where there is no clear external frame of reference. The first categorisation is about the degree of uncertainty involved. What sort of problem is it? High uncertainty will require some sort of planning activity, low uncertainty some form of putting plans into effect. The second dichotomy relates to where the main focus of the problem is. Is it more to do with people, or more to do with structures, organisations, i.e., the "place" where some system is based? These categorisations and the language associated with them are very general, and are applicable to many different

Figure 1: The Four General Kinds of Activity

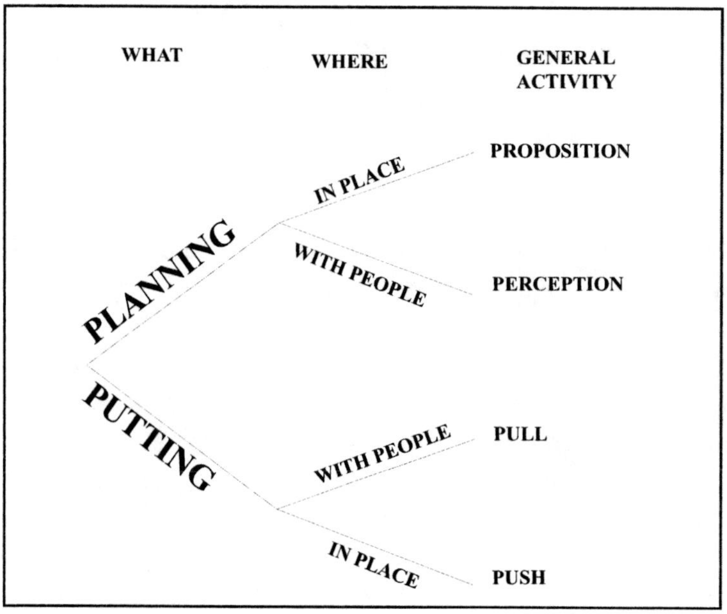

situations. The fundamental generic set of adjustment activities is shown in Figure 1. There are numerous examples of adjustment in management based on these general activities (Brugha, 1998a) and on eight particular activities (Brugha, 1998b).

A most important case is where the decision-maker "owns" the process in the sense that he or she decides "subjectively" when to proceed between stages, rather than when it is in some sense "objectively" "right." The key difference is that, consequently, the decision-maker cannot "pull" himself or herself to make the decision. The tone of the decision then changes from adjustment to "development," in the sense that the decision-maker develops a subjective view or decision. The emphasis then becomes more one of building on levels rather than finding a balance between different activities. Development decision-making can be introverted or extroverted (Brugha, 1998c). The first introverted level is the *somatic*, and refers to tangible things such as needs. Then there are *psychic* (psychological) aspects such as preferences. Finally the *pneumatic* level refers to values or higher goals corresponding to the highest introverted level of *commitment* of one's will. Soma, psyche, and pneuma come from the Greek words for body, soul, and spirit (literally wind).

The extroverted dimension corresponds to stages of *convincing* and starts with technical or self-orientated issues. Then it relates to the context of the problem and how other people see it. Finally it takes account of situations and how to achieve goals or business-purposes. The introverted and the extroverted combine as two dimensions and lead to the construction of nine levels, stages of activity, and types of thinking, and the reconstruction of Maslow's (1987) hierarchy of needs and Jung's (1971) orienting functions (Figure 2).

Figure 2: Levels of Developmental Activities and Types of Thinking

		Convincing Stages		
		Technical Self	Contextual End-User	Situational Goals
Committing Phases	Situational Have/Need	Physical/ Intuiting	Political/ Recognising	Economic/ Believing
	Psychic Do/Prefer	Social/ Sensing	Cultural/ Learning	Emotional/ Trusting
	Pneumatic Are/Value	Artistic/ Experiencing	Religious/ Understanding	Mystical/ Realising

Table 1: Enneagram Numbers, Avoidances, and Traps

Introverted Orientation		Extroverted Orientation	
	Self	*Others*	*Goals*
Somatic	1. Anger/Perfection	2. Need/Service	3. Failure/Efficiency
Psychic	5. Emptiness/ Knowledge	6. Deviance/Security	7. Pain/Idealism
Pneumatic	4. Ordinariness/ Authenticity	8. Weakness/Justice	9. Conflict/ Self-Abasement

Table 2: Systems Development Life Cycle Activities

Introverted Orientation		Extroverted Orientation	
	Technical	**Contextual**	**Situational**
Somatic	Survey project scope and feasibility	Study current system	Define the end-user's requirements
Psychic	Select a feasible solution from candidate solutions	Design the new system	Acquire computer hardware and software
Pneumatic	Construct the new system	Deliver the new system	Maintain and improve the system

Jung's insight was to identify intuiting and sensing as not only different personality types, but also as related to introverted and extroverted dimensions. He named two others as thinking and feeling. These are expanded to seven in Figure 1. The Myers-Briggs test also extends Jung's types, but retains its either-or dichotomies. The Enneagram (Table 1) arrives at the same set of nine types but in a complementary manner. A complication is that Enneagram Type Four needs to be repositioned in order to show a parallel with the other systems (Brugha, 1998c). The differences between Myers-Briggs and the Enneagram are so great that people could learn from doing both tests.

The Systems Development Life Cycle fits this nine-phase structure of convincing within committing (Table 2). Nomology implies that these nine-stage models show different facets of the same structure. Consequently they can inform each other. For instance, the acquisition stage is likely to involve emotions, to require trust, and often be a painful choice. The information systems manager who is attuned to emotions and personality types uses such information in a therapeutic sense and when building teams.

A central claim of Nomology is that adjusting, convincing, and committing comprehensively describe the three dimensions of how the mind structures decisions. For instance, each stage of the Systems Development Life Cycle is carried out as an adjustment process (Brugha, 2001). Consequently, the decision to move from one phase to the next and, within each phase from one stage to the next, is subjective. However, each stage, itself, must be done correctly. So, there are objective ways to decide if one has a good study or design.

Nomology uses a systematic approach to explore constructs that we think we understand. For instance, it suggests that to explore a term such as *spirit*, one should start by asking, "Spirit as distinct from what?" The answers might be *body* and *soul*. Doing it again with the word *body* might suggest the word *mind*. Part of the nomological exploration of "body, mind, soul, and spirit" is to relate them to the generic terms proposition, perception, pull, and push (Figure 1). If they relate properly, there should be a consistent, qualitative difference between each corresponding pair. This should reflect itself in meaningful phrases linking each pair such as "we propose in the body," "we develop perceptions in the mind," "we are pulled in our soul," and "we push with our spirit."

A simpler version of this structure, which is used for personal development, has two levels: committing and adjusting, making a 12-step programme (Peace Pilgrim, 1981). The committing phases are (somatic) preparations, (psychic) purifications, and (pneumatic) relinquishments. Each phase has four steps corresponding to body, mind, soul, and spirit.

A decision structure, where the dominating issue is the wish to adjust, is the Twelve-Step Programme of Alcoholics Anonymous (Anonymous Authors, 1955). Each adjustment phase has three commitment stages. This version can also be used to change people to a higher level of spiritual activity, viz. the Spiritual Exercises of St. Ignatius of Loyola, which has been presented (Tyrrell, 1982; Fessard, 1956) as to "reform the deformed," "conform the reformed," "confirm the conformed," and "transform the confirmed." Clearly these also take place within the areas of body, mind, soul, and spirit. Such a representation could be applied to any adjustment process, depending on how broadly one interpreted the idea of being "deformed."

REFERENCES

Anonymous Authors. (1955). *Alcoholics Anonymous* (2nd edition). New York: Alcoholics Anonymous World Services.

Brugha, C. (1998a). The structure of qualitative decision making. *European Journal of Operational Research, 104*(1), 46-62.

Brugha, C. (1998b). The structure of adjustment decision making. *European Journal of Operational Research, 104*(1), 63-76.

Brugha, C. (1998c). The structure of development decision making. *European Journal of Operational Research, 104*(1), 77-92.

Brugha, C. (2001). A decision-science based discussion of the systems development life cycle in information systems. *Information Systems Frontiers, Special Issue on Philosophical Reasoning in Information Systems Research, 3*(1), 91-105.

Fessard, G. (1956). *La dialectique des excercises spirituels de Saint Ignace de Loyola* (1st edition, pp. 40-41). Paris: Aubier.

Hamilton, W. (1877). *Lectures on metaphysics, Vols. 1 and 2* (6th edition). In *Lectures on metaphysics and logic.* Edinburgh and London: William Blackwood and Sons.

Jung, C. (1971). Psychological types. In *The Collected Works of C. J. Jung, Volume 6.* London: Routledge & Kegan Paul.

Maslow, A. (1987). *Motivation and personality.* New York: Harper & Row.

Peace Pilgrim. (1981). *Steps toward inner peace.* Hemet, CA: Friends of Peace Pilgrim. Available online at: http://www.peacepilgrim.net.

Tyrrell, B.J. (1982). *Christotherapy II.* New York: Paulist Press.

About the Author

Eugene Kaluzniacky is an instructor in the Department of Applied Computer Science and Administrative Studies at the University of Winnipeg in Canada. He has a varied academic, professional and personal interest background that comprises mathematics, statistics, computer science, management science, accounting, information systems, personality psychology, wholistic health, spiritual development and education. He has taught at the undergraduate, graduate, and continuing education university levels and has delivered short courses for government and a community college. A member of the Association for Psychological Type, Eugene has carried out research and consulting on applying the Myers-Briggs Personality Type in IT organizations, on stress in the IT profession, and on IT education. He has also developed a workshop on "Personal Wellness for the IS Professional"—attendees have appreciated his genuineness and enthusiasm. As well, he was a co-creator of an "IT Wellness website," currently at http://itwellness.ncf.ca/. Eugene has co-authored a book on Xbase Programming and has also carried out considerable editorial work. Currently, he is interested in stress management and personal growth in the IT field and has lectured internationally on this topic.

According to Eugene Kaluzniacky, now is the time for the IT professional to broaden his psychological awareness so as to make use of untapped inner energies for a more effective and fulfilling professional and personal life. Eugene is interested in cohesive subgroups within IS organizations that would like to take the ideas in this book "one step further." For example, a group of developers along with its project manager may wish to spend several months systematically documenting the impact of Myers-Briggs awareness on specific IS development tasks. This could enable the eventual development of a full-scale Psychological Factors Methodology. Anyone interested can contact Eugene at e.kaluzniacky@uwinnipeg.ca

Index

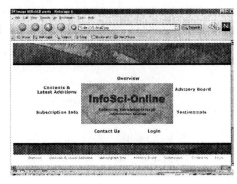